DESIGN OF WIRELESS AUTONOMOUS DATALOGGER IC'S

THE KLUWER INTERNATIONAL SERIES IN ENGINEERING AND COMPUTER SCIENCE

ANALOG CIRCUITS AND SIGNAL PROCESSING
Consulting Editor*: Mohammed Ismail. *Ohio State University

Related Titles:

LNA-ESD CO-DESIGN FOR FULLY INTEGRATED CMOS WIRELESS RECEIVERS
Leroux and Steyaert
Vol. 843, ISBN: 1-4020-3190-4
SYSTEMATIC MODELING AND ANALYSIS OF TELECOM FRONTENDS AND THEIR BUILDING BLOCKS
Vanassche, Gielen, Sansen
Vol. 842, ISBN: 1-4020-3173-4
LOW-POWER DEEP SUB-MICRON CMOS LOGIC SUB-THRESHOLD CURRENT REDUCTION
van der Meer, van Staveren, van Roermund
Vol. 841, ISBN: 1-4020-2848-2
WIDEBAND LOW NOISE AMPLIFIERS EXPLOITING THERMAL NOISE CANCELLATION
Bruccoleri, Klumperink, Nauta
Vol. 840, ISBN: 1-4020-3187-4
SYSTEMATIC DESIGN OF SIGMA-DELTA ANALOG-TO-DIGITAL CONVERTERS
Bajdechi and Huijsing
Vol. 768, ISBN: 1-4020-7945-1
OPERATIONAL AMPLIFIER SPEED AND ACCURACY IMPROVEMENT
Ivanov and Filanovsky
Vol. 763, ISBN: 1-4020-7772-6
STATIC AND DYNAMIC PERFORMANCE LIMITATIONS FOR HIGH SPEED D/A CONVERTERS
van den Bosch, Steyaert and Sansen
Vol. 761, ISBN: 1-4020-7761-0
DESIGN AND ANALYSIS OF HIGH EFFICIENCY LINE DRIVERS FOR Xdsl
Piessens and Steyaert
Vol. 759, ISBN: 1-4020-7727-0
LOW POWER ANALOG CMOS FOR CARDIAC PACEMAKERS
Silveira and Flandre
Vol. 758, ISBN: 1-4020-7719-X
MIXED-SIGNAL LAYOUT GENERATION CONCEPTS
Lin, van Roermund, Leenaerts
Vol. 751, ISBN: 1-4020-7598-7
HIGH-FREQUENCY OSCILLATOR DESIGN FOR INTEGRATED TRANSCEIVERS
Van der Tang, Kasperkovitz and van Roermund
Vol. 748, ISBN: 1-4020-7564-2
CMOS INTEGRATION OF ANALOG CIRCUITS FOR HIGH DATA RATE TRANSMITTERS
DeRanter and Steyaert
Vol. 747, ISBN: 1-4020-7545-6
SYSTEMATIC DESIGN OF ANALOG IP BLOCKS
Vandenbussche and Gielen
Vol. 738, ISBN: 1-4020-7471-9
SYSTEMATIC DESIGN OF ANALOG IP BLOCKS
Cheung and Luong
Vol. 737, ISBN: 1-4020-7466-2
LOW-VOLTAGE CMOS LOG COMPANDING ANALOG DESIGN
Serra-Graells, Rueda and Huertas
Vol. 733, ISBN: 1-4020-7445-X
CIRCUIT DESIGN FOR WIRELESS COMMUNICATIONS
Pun, Franca and Leme
Vol. 728, ISBN: 1-4020-7415-8
DESIGN OF LOW-PHASE CMOS FRACTIONAL-N SYNTHESIZERS
DeMuer and Steyaert
Vol. 724, ISBN: 1-4020-7387-9
MODULAR LOW-POWER, HIGH SPEED CMOS ANALOG-TO-DIGITAL CONVERTER FOR EMBEDDED SYSTEMS
Lin, Kemna and Hosticka
Vol. 722, ISBN: 1-4020-7380-1

Design of Wireless Autonomous Datalogger IC's

by

WIM CLAES

ICsense NV,
Heverlee, Belgium

WILLY SANSEN

Katholieke Universiteit Leuven,
Heverlee, Belgium

and

ROBERT PUERS

Katholieke Universiteit Leuven,
Heverlee, Belgium

Springer

A C.I.P. Catalogue record for this book is available from the Library of Congress.

ISBN 1-4020-3208-0 (HB)
ISBN 1-4020-3209-9 (e-book)

Published by Springer,
P.O. Box 17, 3300 AA Dordrecht, The Netherlands.

Sold and distributed in North, Central and South America
by Springer,
101 Philip Drive, Norwell, MA 02061, U.S.A.

In all other countries, sold and distributed
by Springer,
P.O. Box 322, 3300 AH Dordrecht, The Netherlands.

Printed on acid-free paper

Printed in the Netherlands.

Abstract

The progress in micro-electronics made during the last decades has made the development of highly-intelligent implantable dataloggers feasible. In this work an autonomous miniaturized intelligent datalogger, part of a dental prosthesis, for stress monitoring in oral implants is presented. It monitors the loads on the implants that support the prosthesis in order to gain more insight in bone remodeling processes and implant failures. The datalogger allows to carry out measurements without inconvenience for the patient in everyday living conditions, independent of the hospital environment. It is able to monitor autonomously over a 2-day period, operated by two 1.55-V 41-mAh batteries, so that also unconscious nocturnal dental activities, seen as a missing link in the validation of existing bone remodeling models, can be monitored.

In order to measure the loads, the abutments which are positioned in the gums on top of the oral implants are equipped with 3 strain gauges. By combining the resistance values of the strain gauges the axial force and the bending moment imposed on each abutment can be derived.

The datalogger measures up to 18 strain gauges with an accuracy of 10 μstrain at a sample rate of 111 Hz per channel. It consists of 4 major parts: a sensor interface, a digital part and a bi-directional wireless transceiver, integrated on a single chip, and an external 2-Mbit SRAM memory. The sensor interface includes a reference current source, an 8-bit DAC, a digital interface and compensation memory, a CDS SC amplifier, a CDS SC S/H, a 9-bit successive approximation ADC and a 6-bit programmable relaxation clock oscillator. It processes and digitizes the signals of the strain gauges which are implemented in a current-driven Wheatstone configuration to limit the power consumption. The offset in every channel can be digitally compensated to cope with unwanted offsets due to the strain-gauge resistance tolerance and potential pre-strains.

The digital part, implemented by means of a custom-designed 23.4-kgates FSM consuming 150 μW, orchestrates the operation of the device and increases the intelligence of the datalogger. An automatic-compensation block which performs automatic nulling towards a user-definable output value for a selectable strain-gauge channel is included. Moreover, by the inclusion of an onboard programmable data processing unit with 8 selectable algorithms and adjustable parameters the required data storage capacity is drastically reduced. Only clinical relevant data are stored in the memory, and moreover the data processing can be optimized towards each patient/application.

The bi-directional transceiver allows wireless retrieval of collected data and status bytes from the datalogger, and to reconfigure the measurement device *in situ* after placement of the prosthesis. It is able to communicate over a distance of 30 cm at a data rate of 4 kbytes/s with a mean power consumption of 2.3 mW.

The datalogger IC has been fabricated in a 0.7-μm CMOS technology. The maximum measured mean power consumption of the complete datalogger in its monitoring mode is restricted to 136 μW per strain-gauge channel.

This work demonstrates the feasibility of the single-chip integration of an intelligent strain-gauge datalogger IC, combining a sensor interface with digitally-programmable offset-compensation, a digital unit with adjustable data-processing and automatic offset-compensation, and a wireless bi-directional transceiver. The introduced concepts are not restricted to the presented datalogger alone, but can be applied to a wide variety of portable personal health monitoring systems.

List of Abbreviations and Symbols

Abbreviations

ADC	Analog-to-Digital Converter
AM	Amplitude Modulation
AMP	Amplifi er
ASIC	Application Specifi c Integrated Circuit
ASK	Amplitude Shift Keying
BUF	Buffer
BW	Bandwidth
CCO	Current-Controlled Oscillator
CDS	Correlated Double Sampling
CHS	Chopper Stabilization
CLOCK	Relaxation Clock Oscillator
CMOS	Complementary Metal Oxide Semiconductor
CMRR	Common Mode Rejection Ratio
CNTR	Counter
CRC	Cyclic Redundancy Check
CT	Computed Tomography
DAC	Digital-to-Analog Converter
DC	Direct Current
DFF	D-FlipFlop
DNL	Differential Non-Linearity
EXOR	Exclusive OR
FAME	Food and Animal Monitoring Expert
FPGA	Field Programmable Gate Array
FSK	Frequency Shift Keying
FSM	Finite State Machine
HF	High Frequency
IO	Input/Output
IC	Integrated Circuit
INL	Integral Non-Linearity
ISM	Industrial-Scientifi c-Medical
ISO	International Organization for Standardization
LED	Light-Emitting Diode
LF	Low Frequency
LSB	Least Signifi cant Bit
MEMS	MicroElectroMechanical Systems

MF	Medium Frequency
MSB	Most Signifi cant Bit
MUX	Multiplexer
NAND	Negative AND
NMOS	N-channel Metal Oxide Semiconductor
NRZ	Non-Return to Zero
NTC	Negative Temperature Coeffi cient
OPAMP	Operational Amplifi er
OTA	Operational Transconductance Amplifi er
OTP	One Time Programmable
PC	Personal Computer
PCB	Printed Circuit Board
PGA	Pin Grid Array
PM	Phase Modulation
PMOS	P-channel Metal Oxide Semiconductor
POR	Power-On-Reset
PROG/SEL	Digital interface
PROM	Programmable Read Only Memory
PSD	Power Spectral Density
PSK	Phase Shift Keying
PSRR	Power Supply Rejection Ratio
PTAT	Proportional To Absolute Temperature
PVC	Polyvinyl Chloride
PZT	Lead Zirconate Titanate
RAM	Random Access Memory
RECT	Rectifi er
REG	Nulling memory
RF	Radio Frequency
RFID	Radio Frequency Identifi cation
S/H	Sample-and-Hold
SC	Switched-Capacitor
SIC	Sensor Interface Chip
SMD	Surface-Mount Device
SNDR	Signal-to-Noise-and-Distortion Ratio
SR	Set-Reset
SRAM	Static RAM
SRD	Short Range Devices
STC	Self Temperature Compensated
STSOP	Shrinked Thin Small Outline Package
TCR	Temperature Coeffi cient of Resistance
TTU	Telemetric Temperature Unit
VHDL	Very high speed integrated circuit Hardware Description Language
WDT	WatchDog Timer
ZIF	Zero Insertion Force

Symbols

α	Thermal coeffi cient of resistance
β	Current gain of bipolar transistor
γ	Excess noise factor
Δ	Difference
	Period
$\Delta_{nulling}$	Nulling accuracy interval
$\Delta R_{F_{ax}}$	Resistance change due to axial force
ΔR_{M_x}	Resistance change due to bending moment around the X-axis
$\Delta R_{M_{x'}}$	Resistance change due to M_x with the same proportionality constant as ΔR_{M_y}
ΔR_{M_y}	Resistance change due to bending moment around the Y-axis
ΔV_{DAC}	Voltage step between two consecutive codes of DAC
$\Delta V_{DAC,range}$	DAC voltage range
ε	Strain
ε_{app}	Apparent strain
$\varepsilon_{err,\sigma}$	Measurement accuracy
$\varepsilon_{max,min}$	Maximum/minimum strain
ε_s	Static error
$\phi_{i(d)}$	2-kHz non-overlapping (delayed) bi-phasic clocks
$\phi_{fi\ (d)}$	64-kHz non-overlapping (delayed) bi-phasic clocks
ϕ_{sample}	Special clock
ϕ_{timer}	Timer clock
ϕ_{timing}	Clock used for timing of transmission unit
λ	Linear expansion coeffi cient
ρ	Resistivity
σ	Stress
	Standard deviation
σ_C/C	Mismatch of unit capacitors
σ_I/I	Mismatch of unit current sources
σ_{Mi}	Offset voltage of two matched transistors
σ_{offset}	Input-referred offset
τ	Linear settling time
	Time constant
A	Area
	OTA gain
A_β, $A_{(1)VT(0)}$	Mismatch proportionality constants
A_{AMP}	Amplifi er gain
A_{pre}	Gain of the comparator preamplifi er
b_i	ADC bit number i
BW_{eq}	Equivalent bandwidth
C	Capacitance
C'_L	Total parasitic capacitance at the OTA output
C_{PAR}	Parasitic capacitance
$\overline{CLOCK}$	Clock of the digital part

d	Diameter
	Distance
D	Separation distance
d_i	DAC bit number i
$\overline{di^2}$	Power spectral density noise current
DNL_{max}	Maximum positive DNL-error of DAC
$DNL_{sys,Max}$	Maximum systematic DNL-error
$\overline{dv^2}$	Power spectral density noise voltage
$E_{titanium}$	Modulus of Young for titanium
E_{p-p}	Worst-case peak-to-peak error
$endflag$	Flag to signal end of Analog-to-Digital conversion
f	Frequency
F_{ax}	Axial force
f_{chop}	Chopping frequency
f_{dc}	Feedback factor
G	Gauge factor
g_m	Transconductance
g_{mb}	Bulk transconductance
g_o	Transistor small-signal output conductance
HD_2, HD_3	Second and third-order harmonic distortion terms
I	Current
	Second moment of inertia
I_{DAC}	Digitally-controllable compensation current
$I_{DAC,unit}$	DAC unit-current
I_{REF}	Reference-resistor current
I_S	Saturation current
I_{SG}	Strain-gauge current
I_{SOURCE}	Reference current source
I_{SR}	Maximum output current OTA
I_{unit}	Unit-transistor current
in_i, $prog$, sel_i	Inputs of digital interface
$INL_{sys,Max}$	Maximum systematic INL-error
k	Boltzmann constant
KP	Transconductance parameter MOS
l	Length/height
L	MOS channel length
M	Bending moment
Mi	MOS transistor number i
M_x	Bending moment around the X-axis
M_y	Bending moment around the Y-axis
N_{tot}	Total integrated noise power
p_{cl}	Closed-loop pole
p_{ol}	Open-loop pole
q	Elementary charge
$Q^{+/-}$	Charge at the positive/negative OTA input

Qi	Bipolar transistor number i
r	Radius
R	Resistance
R_0	Nominal strain-gauge resistance
$R_0(S_i)$	Nominal resistance of S_i
$R_{0,eq,Max/Min}(S_i)$	Equivalent worst-case maximum/minimum nominal resistance of S_i
R_{AD}	ADC reference current source resistor
R_{CL}	Oscillator reference current source resistor
r_{ds}	Transistor small-signal drain-source resistance
$R_{H/L}$	ADC reference resistors
r_o	Transistor small-signal output resistance
R_{REF}	Reference resistor
R_x	Receiver
$R(S_i)$	Resistance of S_i
S_β, S_{VT}	Worst-case process proportionality parameters
S_i	Strain gauge number i
SHD_i	Ratio of the signal to the ith harmonic distortion component
sw, SW	Switch
T	Period/time
	Absolute temperature
t_{AD}	Unit settling time ADC
T_{osc}	Oscillator clock period
t_{reg}	Regeneration time
t_s	Total settling time
t_{SR}	Slewing time
T_x	Transmitter
T(s)	Transfert function
TC	Temperature coeffi cient
TCL	Linear temperature coeffi cient
TCQ	Quadratic temperature coeffi cient
V	Voltage
V_{acc}	Input-reffered minimum accuracy level
V_{BE}	Base-emitter voltage
V_c	Comparator input voltage
V_{DD}	Supply voltage (3.1 V)
V_{DS}	Drain-source voltage
V_{DSsat}	Saturation voltage
$V_{err,\sigma}$	Amplifi er-output-referred voltage accuracy-level
$V_{err,\sigma,in}$	Input-referred voltage accuracy-level
V_{f,ϕ_i}	Voltage at the end of the phase
V_{GS}	Gate-source voltage
V_{GST}	Gate-source overdrive voltage, i.e. V_{GS}-V_T
V_H	High reference voltage of ADC
V_{i,ϕ_i}	Voltage at the beginning of the phase
V_L	Low reference voltage of ADC

V_{MM}	Intermediate voltage (1.55 V)
V_{off}	Offset voltage
V_{quant}	Quantization noise
V_{ref}	Reference voltage of ADC
$V_{rms,in}$	Input-referred rms noise voltage
$V_{S,low/high}$	Lower and upper threshold of Schmitt trigger
V_{SB}	Source-bulk voltage
V_{SS}	Ground (0 V)
V_T	Threshold voltage
	Thermal voltage
W	MOS channel width
Y	Yield

Contents

Chapter 1

Introduction

The **objective** of this work is the development of an autonomous miniaturized intelligent datalogger, embedded within a dental prosthesis, for stress monitoring in oral implants. The device is used to investigate the loads acting on the implants, supporting the prosthesis, by means of strain gauges. The goal of this research is to gain more insight in the processes involved in bone remodeling and implant failures. The datalogger is able to measure up to 18 strain-gauge channels and features automatic offset-compensation for each channel and programmable on-board data processing, both implemented in its digital part. A wireless bi-directional transceiver is also included, which allows to reconfigure the device *in situ* after placement and to retrieve the collected data and status bytes. Because the datalogger is battery-operated, special care has been taken to restrict its power consumption.

The **outline** of the presented work is as follows:

- In Chapter 2 an overview of the most important design aspects of miniaturized biotelemetry systems is given. The different building blocks are discussed and two implantable dataloggers are presented as an example.

- Chapter 3 introduces the system concept of the new miniaturized datalogger. The clinical background and the motivation behind the presented work are given. Furthermore, the measurement methodology and the external non-portable measurement setup, which has been developed first, are discussed. A comparison between semiconductor and metal film strain gauge sensors is made and the datalogger's specifications are described.

- Due to its complexity the realization of the datalogger has been carried out in two stages. First a separate sensor interface chip has been developed. This chip is described in Chapter 4. First the implemented measurement setup and offset-compensation setup are presented. The design of the different building blocks of the sensor interface is discussed next. After illustrating the layout of the realized chip, the chapter ends with experimental results.

- Chapter 5 describes the development of a single-chip intelligent-datalogger IC, which combines the sensor interface, the digital part and the wireless transceiver. The operation principle of the transponder-type datalogger, which also includes an external RAM, is explained, followed by a digression of its digital part and transceiver. The automatic offset-compensation and the programmable data processing unit are elaborated. After that, the realized chip and its measurement results are presented. The chapter ends with a first concept study of the datalogger's packaging.

- Finally, Chapter 6 presents some general conclusions.

Chapter 2

General design aspects of miniaturized low-power dataloggers

2.1 Introduction

During the past three decades a tremendous progress has been made in electronic systems for medical applications. They have progressed from discrete-transistor devices to implantable highly-intelligent integrated sensor-systems [Wou 95, Wis 94, Mok 99], merging sensors, actuators, analog interface circuits, digital intelligence and telemetry. A commonly-known example of such an implantable system is the cardiac pacemaker [Wer 00]. Today, implantable cardiac pacemakers are used with a high success rate as a long-term, safe and reliable form of therapy for different kinds of cardiac disrhythmia. Different technological developments, such as highly-integrated circuits and the use of lithium batteries, have been milestones in the development of pacemakers. During their history, beginning with the first implantation in 1958 (Fig. 2.1 (a)), a technological revolution has occurred. The early devices, consisting of a few transistors, were only able to pace with impulses operating at a constant rate and amplitude, and had a short lifetime. Modern, rate-adaptive dual-chamber pacemakers (Fig. 2.1 (b)) have a weight of only 25 g and are getting increasingly smaller. They possess highly-complex integrated circuits and can pace the right atrium and the ventricle, monitor the intrinsic cardiac activity, adapt automatically to changing needs of the heart, be adjusted through inductive telemetry, and guarantee a lifetime of 8 years and longer. Research is also going on to integrate an accelerometer into them in order to monitor the patient's physical activity and adapt the operation of the pacemaker automatically to this. This example clearly illustrates that a lot of progress has been made in the field of sensor systems during the last decades.

In this chapter an overview of the most important design aspects of sensor systems, more in particular, of miniaturized biotelemetry systems, is given. The design considerations introduced in this chapter also apply to the datalogger used for stress monitoring in oral implants, which is presented in the following chapters.

First, a definition of biotelemetry systems is given. The most important design criteria are highlighted and the advantage of establishing a bi-directional communication link is explained.

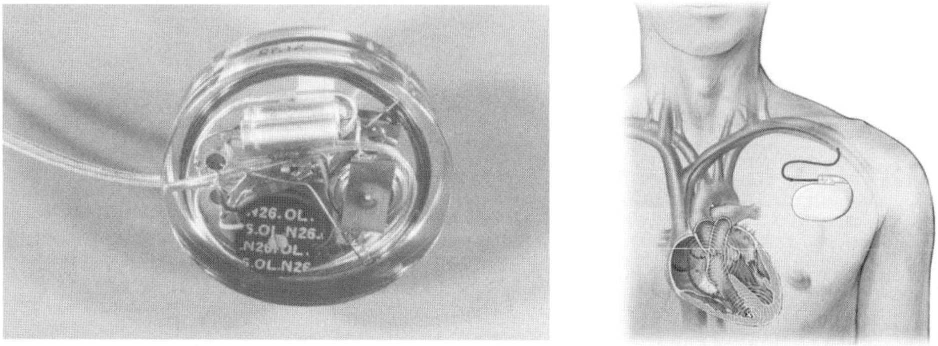

(a) (b)

Figure 2.1: *(a) The first implanted pacemaker and (b) a modern dual-chamber pacemaker (Guidant).*

Next, the general architecture of an implantable datalogger is described. It includes sensors, signal conditioning, data processing, a transceiver, memory and a power source. The design choices regarding these building blocks and the packaging of the datalogger are discussed. Also the concept of a smart sensor system is presented. To conclude two examples of injectable telemetric dataloggers are given. The first device is an injectable datalogger used for animal identification and quantification of animal welfare. The second device is a redesigned version of the first one, intended for commercial use.

2.2 Biotelemetry systems

Biotelemetry, literally *measurement of biological parameters from afar*, is defined as the measurement and transmission of biomedical/physiological parameters from an often inaccessible location (e.g. implantable systems) to a remote receiver site, thereby inducing as less stress or discomfort as possible for the subject under surveillance by the monitoring itself [Wou 95]. Only when the human/animal under surveillance is not hampered or stressed by the monitoring equipment in any way, and is allowed to conduct its normal behavior and daily activities, relevant and accurate parameters can be monitored. The possibility to measure continuously over a longer period in normal living conditions without any hindrance (e.g. due to wires/cables) for the subject is a major advantage of (implanted) biotelemetry systems compared to conventional measurement systems connecting the sensors with wires.

The most important design issues for implantable biotelemetry systems are: *low power consumption*, resulting in a long life time for battery-operated systems; *miniaturization and low weight*, making implantation without nuisance for the subject under surveillance feasible; *packaging*, protecting the electronics from the surrounding body fluids [Pue 96]; and *high reliability*, because the system can not be repaired after implantation. After implantation of these moni-

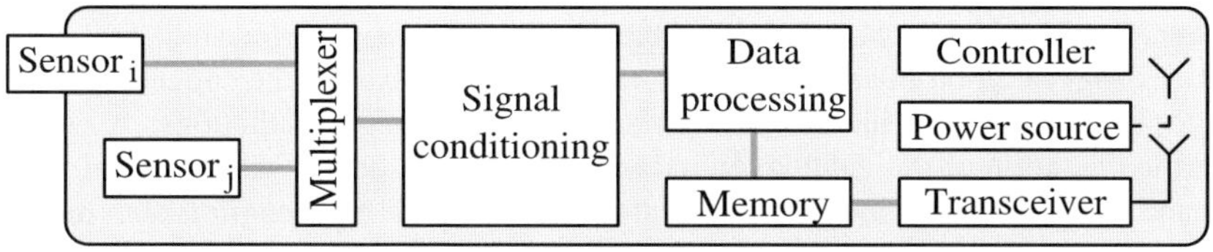

Figure 2.2: *Overview of a datalogger. Actuator(s) are not shown.*

toring systems in the body of an animal/human, they are inaccessible for repair, replacement or adjustments, unless a surgical intervention takes place. During recent years, the need for remotely-adjustable systems has lead to the development of biotelemetry devices with read *and* write facilities. To achieve this they are equipped with a transceiver instead of with a transmitter only. The resulting bi-directional link allows not only to read data *from* the monitoring device, but also to send commands *to* the device, so that its configuration can be (re-)programmed. This (re-)programmability is a major advantage for many reasons. By the incorporation of flexibility into the operation of the device during the design phase (e.g. number of channels, sample frequency, data processing algorithm, ...), and foreseeing the possibility to wirelessly (re-)program the device settings *in situ*, a very flexible monitoring device can be realized, which can be adapted to different needs as required by the situation or the application. This (re-)programmability is especially useful in novel applications, where the end user has no experience with measuring on the remote site. The ability to reprogram the sensor unit *after* installation allows to overcome uncertainties at the beginning of the novel research activities.

Another advantage of the implementation of the bi-directional link is the possibility to deal with the long-term drift of sensors [Pue 99]. An ideal sensor should be stable in time, so that its associated sensor interface is able to follow the sensor signals at all times. In real life though most sensors cannot fulfill this requirement. Especially in long-term applications, sensor drift is a commonly-known problem. In many cases, the drift of the sensor becomes so important that its associated amplifier is saturated, resulting in signal loss. The bi-directional communication link reaches a possible solution to deal with this problem. The device can be commanded to go into a self-interrogating mode to verify the actual status of the different building blocks. After transmission of the status data, the remote controlling station can adapt the settings of the system to cope with potential drift. State-of-the-art sensor systems and sensors [Rey 02] go even one step further. They are equipped with a (continuous) built-in self-test [Coz 99, Deb 02] and/or auto-calibration unit [Mei 94]. In this way they are capable to continuously adapt themselves, and they can operate autonomously without any intervention from the outside with regard to sensor drift, enhancing their reliability and operating security.

2.3 Dataloggers

Dataloggers equipped with a wireless telemetric link form an important subdivision of biotelemetry systems. Fig. 2.2 shows a general overview of the most important building blocks of such a datalogger [San 82]. It is capable of storing the measured biomedical/physiological data (mostly after processing) in an on board *memory*. The wireless retrieval of the collected data from the datalogger is done after the measurement interval or at interim consultations. A *controller* supervises the operation of the datalogger and reconfigures the device settings upon reception of a command. The selection between different sensors is done by a *multiplexer*. In addition to sensors, actuators (not shown) can also be included. The building blocks of the datalogger are explained now.

2.3.1 Sensors

Sensors have a growing market potential [Wec 02]. Typical application areas are the automotive industry, consumer electronics, and biological and medical equipment [Bol 95]. The essential task of a *sensor* [Hos 97b] is to convert a signal from one energy domain [Mid 89], i.e.:

- mechanical - force, pressure, velocity, acceleration, position, flow
- thermal - temperature, heat, heat flow
- chemical - concentration, composition, reaction rate
- radiant - electromagnetic wave intensity, phase, wavelength
- magnetic - field intensity, flux density
- electrical - voltage, current, charge, resistance, capacitance, polarization

into an electrical signal, which can be 'conditioned' further by sensor interface electronics.

Ideally, the output of a sensor is proportional to its input signal and remains the same over time when the same input signal is applied. Unfortunately, real sensors drift, have offsets, are non-linear, and their output signal is often noisy and very weak. Note that the latter is especially true for implanted dataloggers equipped with sensors, because of their low-power requirement. Sensors also have cross-sensitivities to parameters other than the measurand of interest, like e.g. temperature or supply voltage, and their leads may pick up interfering noise signals. To cope with the non-ideal behavior of the sensor dedicated functions are implemented in the signal conditioning block: *filtering* to deal with out-of-band interfering noise; *calibration* to cope with the variations in offset and sensitivity, and the non-linearities of an ideally linear sensor; and *compensation* for temperature or for the sensor's non-linear (e.g. logarithmic) characteristic. Note that non-ideal behavior of the sensor system not only can be caused by the sensor, but also by the sensor interface electronics. E.g. a sensor interface with a low input impedance can strongly affect a sensor with a rather high output impedance. One of the tasks of the signal conditioning circuitry is to provide an appropriate impedance conversion to interface with the sensor without affecting the operation of the latter. As explained in [Hos 97a], the deviations from ideal behavior of the sensor and the sensor interface electronics can be classified into *time-variant* and *time-invariant*, and *deterministic* and *statistical* non-idealities.

Well-characterized *time-invariant deterministic* non-idealities, such as well-defined device non-linearities and parasitics, can be taken into account *prior* to fabrication. They can be accounted for during the design phase.

On the contrary, if the parameters of the sensor and/or sensor interface electronics are affected strongly by *time-invariant statistical* variations, e.g. due to manufacturing processes, it is not possible to take these non-idealities into account prior to the fabrication of the sensor system. The effects of such non-idealities, like e.g. variations of the sensor sensitivity and offset, must be dealt with by calibration *after* fabrication. *Calibration* is defined as the process of applying reference signals to the sensor system, which provides the necessary correction parameters to adjust the sensor output signal, so that its input-output relation is known with a certain accuracy. This correction can be carried out on board of the datalogger or on the remote station after retrieval of the data. The calibration procedure should be kept as simple as possible preventing that it increases the total cost of the sensor system significantly [Hor 97]. Note that testing (1/3 of the overall cost), including calibration, and packaging (1/3) are time and money consuming tasks. They have to be taken into account from the start of the sensor system development. The fabrication cost of the sensor system itself is only 1/3 of the overall cost [Hab 97].

In order to be able to *compensate* for *time-variant deterministic* non-idealities continuous monitoring of these variables is required. An example of this class of non-idealities is the cross-sensitivity of the sensor system to power-supply-voltage and temperature variations. The resulting error can be minimized by measuring continuously the interfering signal inducing the error, and compensating for the error appropriately. The compensation for temperature for instance can be done by measuring the temperature with a (on-chip) temperature sensor, resulting in a multi-sensor system, and correcting the sensor-system output depending on the measured temperature. A possible solution to deal with a high cross-sensitivity to temperature of a sensor may be the addition of an extra sensor, identical to the original one, but which is only susceptible to the temperature signal. The original sensor measures the wanted signal (spoiled by the temperature-induced signal), while the extra sensor measures only the temperature-induced signal. By subtraction of the latter from the first signal, the wanted signal, depending ideally only on the measurand of interest, can be derived.

By far the most difficult non-idealities to treat are *time-variant statistical* non-idealities. Examples of these are interfering noise, uncorrelated drift, and aging. Knowledge of the sensor signal in combination with appropriate signal processing algorithms, like e.g. filtering and correlation techniques, can reduce the effects of these non-idealities significantly [Hos 97a]. Also the possibility to (re-)configure the device by the bi-directional link may offer a solution.

2.3.2 Signal conditioning

An important function of the signal conditioning block (Fig. 2.2) is low-noise and low-offset amplification of the weak sensor signals. Generally, it must also be able to adjust the offset and the sensitivity of the sensor in order to ensure that the implemented amplifier is not saturated and the dynamic range of the sensor system is not degraded. The sensitivity adjustment may be performed by adjusting the gain of the amplifier. Furthermore, the conditioning block must offer an appropriate impedance conversion to interface with the sensor without affecting the operation

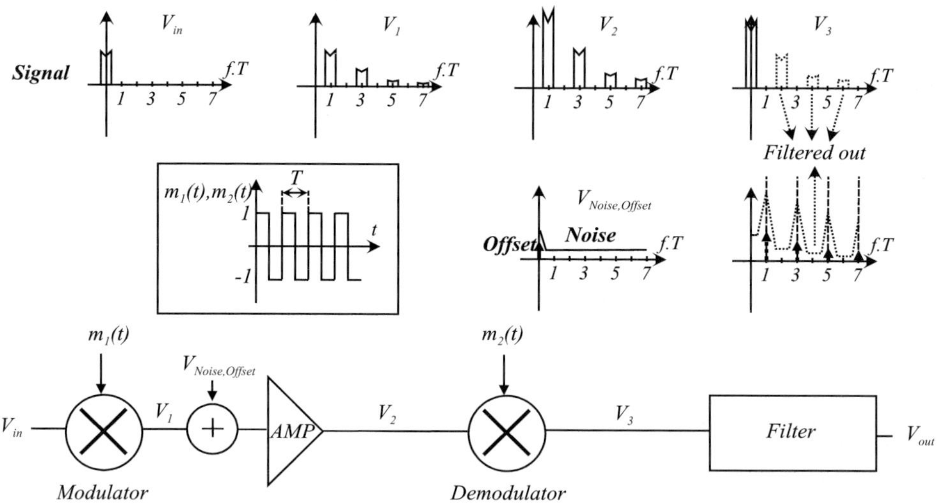

Figure 2.3: *Principle of the chopper amplifier* ($f.T = \frac{f}{f_{chop}}$).

of the latter. Another important function of the conditioning block is analog-to-digital conversion, converting the amplified analog sensor signals into the digital domain. By working in the digital domain, a high interference immunity and robustness against component degradation can be obtained. These properties are important considering the harsh environment to which some sensor systems are exposed. They are the reasons why in some modern sensor systems the sensor is directly implemented in the analog-to-digital converter (ADC), providing the digital signal as soon as possible [Lem 93]. The amplifier and ADC are often complemented with additional signal conditioning functions, such as filtering, calibration and (temperature) compensation. All these conditioning operations prepare the sensor signal for subsequent evaluation by the data processing unit. Note that in most cases a specific conditioning block is required for a given sensor/application. Because of this specific character, generic interfaces, designed to interface a range of sensors for a variety of applications, generally give rise to a large area/power consumption and/or inferior accuracy in comparison with a customly-designed conditioning block for one sensor/application.

As already mentioned above, low-noise and low-offset amplification is an important function of the signal conditioning block. Because sensor interfaces are mostly restricted to low-frequency applications, especially 1/f noise can pose a problem. The in-band 1/f noise of the amplifier may be larger than its in-band white noise. Not only amplifier noise but also amplifier offset can cause troubles, because the signal produced by the sensor generally is of a smaller magnitude than the offset voltage of the amplifier. This is especially true for dataloggers with resistive sensors in applications which require a low power consumption. Moreover, the amplifier offset is dependent on temperature. In long-term measurements, also the drift of this offset voltage can pose a problem. Two possible solutions to cope with the amplifier 1/f noise and offset (drift) are the chopper amplifier (CHS) [Enz 96], and the switched-capacitor amplifier, based on the Correlated Double Sampling (CDS) technique.

The principle of the chopper amplifier [Enz 87] is shown in Fig. 2.3. First, V_{in} is considered. V_{in} is the output signal of the sensor, that needs to be amplified. It has a frequency spectrum, that must be smaller than $f_{chop}/2$. This signal is modulated by the square wave $m_1(t)$ with a period $T=1/f_{chop}$. This modulation transposes the signal's spectrum around the odd harmonics of f_{chop}. It is then amplified and demodulated back to the original band by $m_2(t)$, which is the same as $m_1(t)$, if no phase shift is introduced by the amplifier. Otherwise the phase shift needs to be compensated for. The output signal of the demodulator contains spectral components around the even harmonics of the chopper frequency, which are filtered out by the subsequent filter, so that the amplified 'baseband' sensor signal is obtained at the output V_{out} of the filter, as shown in Fig. 2.3. Next, the (low-frequent 1/f) noise and the offset (drift) $V_{Noise,Offset}$ of the amplifier are considered. These are modulated only once by $m_2(t)$ and their spectrum is translated to the odd harmonics of f_{chop}. They are further filtered out by the subsequent filter such that 'only' the desired amplified sensor signal is obtained, without amplifier 1/f noise and offset (drift). If the chopping frequency is much larger than the noise corner-frequency [San 94], the residual in-band white noise at the output is only very slightly larger than it would have been without chopping [Enz 96]. The problem encountered with chopper amplifiers is the presence of spikes at the input modulator, which are a consequence of charge injection mismatch (cf Section 4.3.4.6). These spikes lead to a residual offset [Enz 87]. To cope with this, an additional bandpass filter [Men 97] before the second modulator and on-chip tuning [Men 98] of this bandpass filter to the oscillator, used to create the chopper frequency f_{chop}, are employed in present-day state-of-the-art chopper amplifiers. The tuning accuracy of f_{chop} to the bandpass resonance frequency is the limiting factor for the residual offset reduction.

The basic idea behind the Correlated Double Sampling technique is the sampling and storing of the offset during one phase and 'subtracting' the sampled offset from the 'new' offset occurring during the next phase. Because the offset variation with temperature and the drift of the offset are slowly-varying signals, the two offset values are strongly correlated, given that the time between the two phases is sufficiently small. By 'subtraction' of the offset values of the two successive phases offset-cancellation is achieved. The CDS principle is not only used to cancel the OTA's offset, but also to reduce the 1/f-noise. Because the low-frequent character of this noise, two subsequent 1/f-noise values are strongly correlated too, so that also the 1/f-noise noise contribution is reduced by this technique. More details of this technique and implementation examples are given in Chapter 4.

An advantage of the chopper amplifier is that the white noise of the amplifier is not aliased into the baseband, contrary to the switched-capacitor amplifier (kT/C noise [San 94]). This suggests that the chopper amplifier is more appropriate for continuous-time applications, whereas the switched-capacitor amplifier is more suitable for sampled-data applications, where aliasing is unavoidable [Enz 87].

A detailed comparison of the relative advantages and disadvantages of the chopper stabilization (CHS) versus CDS technique has been given in [Enz 96]. These are the main conclusions:

- CDS is inherently a sampled-data method. CHS is based on modulation rather than sampling, avoiding aliasing, and hence can be used for continuous-time signals.
- CDS reduces the low-frequency noise by high-pass filtering (cf [Enz 96]). CHS translates it to some out-of-band frequency.
- The output noise of a CDS amplifier is normally dominated by aliased wide band noise resulting from undersampling. In a continuous-time CHS amplifier the noise spectrum is not folded, and hence 1/f noise remains dominant in the baseband *before* the second modulation. *After* demodulation, white noise is transposed into the baseband and replaces the 1/f noise which is translated to the odd harmonics of f_{chop}. If the chopping frequency is much larger than the noise corner-frequency, then the baseband white noise in the output is only very slightly larger than it was without CHS (cf [Enz 96]).
- CDS can also be used to enhance the effective gain of the employed OPAMPs/OTAs [Joh 97]. A continuous-time CHS amplifier, by contrast, causes the OPAMP/OTA to amplify a high frequency signal, and hence its effective gain is usually reduced.

In conclusion, CDS is preferable in applications which inherently use sampled-data circuits (such as Switched-Capacitor circuits), so that the baseband noise behavior is not deteriorated by noise aliasing. Also, the DC offsets are eliminated, not just modulated to a higher frequency, by CDS, which may improve the allowable signal swing. Finally, the gain-enhancing ability of CDS may be an important advantage in some applications. On the other hand, CHS is the method of choice if low baseband noise is an important requirement and if the system is a continuous-time one to start with.

2.3.3 Data processing

The signal conditioning block is followed by a data processing block (Fig. 2.2), which extracts relevant information from the conditioned sensor signal(s). The implemented algorithms may range from simple ones as presented in Chapter 5 to more complex ones, like spectral analysis, signal compression or pattern recognition. Note that both analog and digital circuit techniques may be applied for the implementation of the data processing. An example of the former is given in Section 2.4.1. Usually the data processing is performed on board of the datalogger. For some devices though, mostly passive devices (cf Section 2.3.4), a 'continuous' telemetry link is established and the data processing is carried out on an external processing device/computer.

Because of the extreme miniaturization that is required for implantable dataloggers, only a limited data storage capacity is available on board. By extraction of relevant data from the sensor signals and storing only these in the memory, the required memory capacity can be significantly reduced. Moreover, since the amount of data that needs to be transmitted is reduced, the telemetry link, which generally has a relatively large power consumption, is much more efficient by the implementation of the data processing block.

Usually a (on-chip) microprocessor (or microcontroller) or a dedicated Finite State Machine (FSM) is used to implement the controller (Fig. 2.2) and the (digital) data processing. In general the drawbacks of a microprocessor are its power consumption and size. Even if special low-power modes (sleep, standby) are provided, the mean power consumption is still often too large. In most cases this is intolerable, because implantable telemetric devices are intended to have a long life time. Because in many applications also extreme miniaturization of the datalogger is necessary, the inclusion of a microprocessor may result in a too large volume. Moreover, a lot of the processor capabilities may be redundant for the given application. Therefore, FSMs are often a better solution. Especially for large quantities a dedicated FSM is desirable, because it may reduce the fabrication cost of the total system. Moreover, the FSM has a lower power consumption. On the other hand, the advantages of a microprocessor are the possibility to change the operation of the controller/data processing easily by changing the code, a short design time, which may result in a reduced design cost, and a high system expendability.

The choice between a microprocessor-based or a FSM-based sensor system is determined by the application itself, the development stage, the available time to market, the required design efforts, ... From the technological point of view a microprocessor is the best solution if power consumption and size are not critical and if it is desirable to be able to change the processing unit/controller in the future. In the case where miniaturization, power consumption and cost are critical, a dedicated FSM is probably the best option. It is good practice for both cases though to implement sufficient flexibility, so that the datalogger is still adaptable by the wireless link.

2.3.4 Power source

Implanted devices are intended to have a long life time, restricting the need for surgical interventions as much as possible. In contrast with non-implanted systems, where the batteries can be changed easily at periodic intervals, battery replacements are not possible for implanted systems without additional surgery. Therefore, an extreme low power consumption is required for these devices, which might e.g. be achieved by switching off circuit parts, which are not in use. Moreover, implanted systems also require sufficient miniaturization, restricting the available volume and thus also the maximal capacity of the implemented batteries.

Biotelemetry systems can be divided into two main categories [Wou 95] with respect to their powering:

- **Active devices:** These devices [Pue 96] are equipped with an on board battery. No interaction with the outside world is required regarding their powering, so that they are able to monitor autonomously on a continuous base. They usually have a larger transmission range than passive devices. Two popular battery types for biotelemetry systems are silver oxide and lithium batteries. The former have a very flat discharge characteristic, which is very beneficial for analog design. The latter have a high energy density and a long shelf life, but on the other hand they have a relatively high internal resistance. Therefore they are most suited for low current drain applications.

- **Passive devices:** These devices [Pue 96, Zia 97] derive their power from an external radiofrequent (RF) powering field (usually based on the principle of inductive coupling).

They are only able to operate if this RF-powering field is active *and* in the proximity. The demand for the latter severely limits their operational range. Furthermore, the freedom of movement of the subject under surveillance may be strongly hampered by this requirement. Moreover, the powering fields may also interfere with the weak sensor signals. Advantages of passive devices are a prolonged life time and a possible reduction in volume and weight due to the absence of batteries.

The choice between a passive and an active system depends on the application. A passive device may be selected if a non-continuous monitoring with a short-range communication link is satisfactory *and* if it is feasible to bring the external powering system in close proximity to the monitoring device. RF-powering by a portable energizer may also offer a solution if too much continuous power is required by the sensor system and the battery would otherwise be empty in no time. On the other hand, if continuous monitoring is required and the power consumption is sufficiently low, resulting in a long life time, an active system is preferred. Especially for applications where powering is of vital importance (e.g. pacemakers) only the active approach provides a reliable solution.

RF-powering can also be used to recharge rechargeable batteries employed in implantable devices. In this way the life time of the implanted devices can be extended. However, it must be kept in mind that the number of possible recharges for rechargeable batteries is limited and that regular recharging is required due to their relatively high internal leakage current (self-discharge), which is especially true for NiCd and NiMH rechargeable batteries.

2.3.5 Transceiver

The wireless transceiver eliminates the need for cables to connect to the (implanted) device, making measurements without inconvenience feasible. Biotelemetry systems usually employ radiofrequent (RF) techniques/carriers for their communication, offering solutions for short as well as long-range applications. The selection of an optimal radio frequency for the operation of a given biotelemetry system requires consideration of several factors, including technical performance, regulatory issues, and the interference with other systems. The need to exercise care with regard to other radio services significantly restricts the available range of operating frequencies. Therefore, it is usually only possible to use frequency ranges that have been reserved specifically for industrial, scientific or medical applications or for short range devices (SRD). These are the frequencies classified worldwide as ISM frequency ranges (Industrial-Scientific-Medical) and SRD frequency ranges. Low-frequency ($<$ 135 kHz) systems are generally based on inductive coupling, have a limited communication speed and a short communication range. An advantage of low-frequency radio signals is their ability to propagate through water, body tissue, and through considerable thicknesses of other materials. This makes them very suited for implanted devices. They are e.g. used in injectable animal identification-and-monitoring devices (cf Section 2.4). Another advantage is their limited performance degradation by metallic objects. High-frequency (e.g. 433 MHz/916 MHz) systems on the other hand offer long communication ranges and a high communication speed. Moreover, they allow for the use of smaller antennas. An important drawback of high-frequency radio signals however is their attenuation by many common materials, especially if the moisture content is high. Human tissue e.g. acts as an absorber for frequencies above 100 MHz [Wou 95]. Therefore, high frequencies are not suited for implanted devices. Moreover, they are also reflected by metals, which may also limit their performance. They are used e.g. for railroad car tracking and automated toll collection.

Because the transceiver is a major power consumer on board of the datalogger, its design must be optimized towards a low power consumption. Note that the required miniaturization of the antenna/coil for implanted devices results in a low antenna efficiency, and hence increases the minimum required power consumption. Note also that the maximum power of radio signals in the neighborhood of living tissue is regulated to avoid potential damage [Wou 95].

2.3.6 Packaging

The packaging of the datalogger is a very important and often difficult task to fulfill [Bol 95]. For most applications specific packaging strategies have to be developed. On one hand, the housing has to be transparent for the measurand. On the other hand, the packaging has to shield the sensor and the electronics from unwanted effects such as humidity, dirt, mechanical forces, light, etc. Moreover, for implantable dataloggers biocompatibility is a very important design criterion for the packaging.

The packaging problem depends on the sensing principle [Pue 96]. As depicted in Fig. 2.2, a distinction must be made between devices with sensors ($Sensor_j$), which can be hermetically encapsulated within the package (e.g. accelerometers, temperature sensors, magneto sensitive sensors, etc), and devices with sensors ($Sensor_i$), which inevitably must be exposed to the measurand (e.g. pressure, flow, chemical sensors, etc). It may be clear that the packaging demands in the latter case may become quite severe, and in many cases lead to serious reliability problems, presenting a real barrier towards the applicability of the sensor system [Bow 86]. As a rule of thumb, one should always make an effort to design the system such that 'contactless' sensors, which can be hermetically encapsulated within the package, can be adopted. It is important to take the packaging into account from the beginning of the sensor system development in order to be able to validate its overall performance.

2.3.7 Smart sensors

Most of the sensor systems realized so far consist of discrete sensors combined with one or more ASICs or commercial components, on a printed circuit board (PCB) or a hybrid carrier [Mal 96]. However, the progress recently made in industrial IC technology, mostly combined with extra post-processing steps, has made the realization of miniaturized sensors and MEMS in silicon possible. Besides its electrical properties, silicon exhibits outstanding mechanical properties. Piezoresistive, thermoelectric and magnetic effects are intrinsic to silicon and can be exploited to realize sensors. At present time, it is possible to co-integrate the conditioning-and-processing circuits and the sensors on the same chip. These systems, often equipped with on-chip calibration, a self-test and a standardized bus output, are referred to as smart sensors [Hui 92, Hui 94].

On one hand, smart sensors have several advantages. While a sensor in the traditional sense outputs raw data, a smart sensor outputs 'only' useful information and may be dynamically

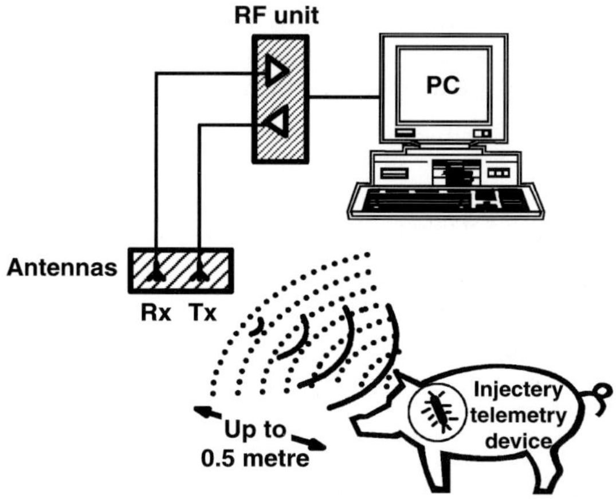

Figure 2.4: *Setup of DEVICE 3.*

programmed as user requirements change. Their cost is strongly reduced, because the use of standard IC technology makes batch fabrication possible. Moreover, the reduction in interconnections (and size) may improve their reliability and performance (less parasitics). The reduced cost in combination with on-chip calibration/self-testing and a standardized bus interface, making it easy to implement/replace them, makes them very attractive for consumer products [Hor 97].

On the other hand, the choice of materials compatible with silicon IC technology for integrated *sensors* is quite limited and their properties are process-dependent. The integration of sensors in an IC process is challenged by the constraints of the fabrication process and the very tight control on material properties required to produce functioning electronic devices with predictable characteristics [Lem 93]. The demands of the sensor and the sensor interface electronics are conflicting. Therefore, integrated sensors are often less performant than their discrete counterparts developed in dedicated technologies, resulting in weak signals, offset, and nonlinear transfer characteristics. They thus put increasing demands on the interface circuits.

The accuracy of the integrated components of the *sensor interface electronics* may also be degraded by the post-processing steps necessary for most integrated sensors. Moreover, in several sensor applications (such as automotive, biomedical, environment monitoring and industrial process control) the smart sensor chip may be exposed to harsh environmental conditions, causing aging and degradation of the on-chip electronic devices. This makes most circuit techniques, which rely on accurate component matching and complex analog functions, in these cases inconvenient. Interface circuit design therefore requires specific knowledge and special techniques in order to achieve the required performance and reliability [Lem 93]. Another drawback of the the co-integration of the sensor and its interface is the interference of e.g. digital signals of the processing unit with the very weak sensor signals [Rie 94]. Moreover, self-heating may also introduce temperature interference.

Contrary to fully integrated smart sensors, the use of a separate sensor and a separate sensor-interface ASIC (i.e. hybrid system), both demanding specific design knowledge, gives more flexibility during the design, and allows separate optimization of both and selection of the optimal technology for both.

2.4 An injectable transponder example: from prototype to commercial device

An excellent introduction to the state-of-the-art (at the beginning of this work) is the autonomous datalogger developed in [Wou 95]. The prototype of this device, called DEVICE 3, has been developed within the MICAS group [Wou 95]. It is a miniaturized injectable biotelemetry transponder prototype, used for identification and measurement of temperature and activity in large scale animal husbandry. Still unfinished in 1996 it has been further improved to show feasibility of the concept. The second device, called DEVICE 4, is a redesigned version of DEVICE 3, intended for commercial use in the animal husbandry and the food distribution sector. The design of the second device has been done by BRUCO [Bru] in close collaboration with MICAS.

2.4.1 Prototype development: DEVICE 3

2.4.1.1 General overview

As already presented in [Wou 95], DEVICE 3 has been developed to quantify animal welfare, which is related to temperature and activity. The (autonomous) device, powered by a small lithium battery, measures temperature and movement data by means of a thermistor and two accelerometers and stores the data in an on-board memory. When the device is wirelessly activated by an external field (transponder principle), the stored data or the identification code of the device, dependent on the (programmable) operation mode, are transmitted to an external transceiver/PC for further processing and visualization. The total setup of the device is shown in Fig. 2.4 and a detailed block diagram of the device itself is depicted in Fig. 2.5.

Besides the sensors, the device consists of three major parts: a Sensor Interface Chip (SIC), a microcontroller, that incorporates intelligence in the transponder, and a transceiver. The presence of the wireless bi-directional communication link is a great advantage. In this way not only data can be read from the device, but also commands can be sent to the device to program its settings. As a result, the measurement algorithms for temperature and movement (cf Table 2.1) and other device settings can be optimized towards the application. The measurement system may also be adjusted to cope with long term drift of the sensors and/or electronics. The commands that can be issued to the SIC are listed in Table 2.1.

2.4.1.2 Modes of operation

DEVICE 3 has three modes of operation:

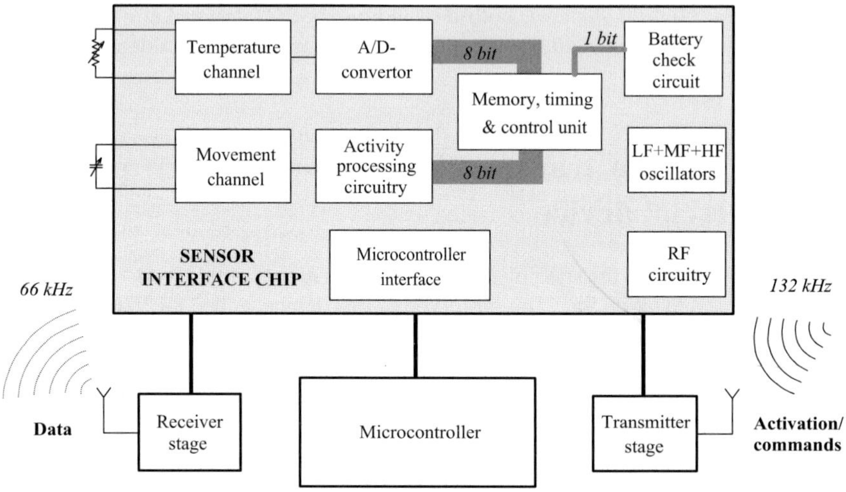

Figure 2.5: *Overview of the building blocks of DEVICE 3.*

Command	Action
0000	Stop and general reset
0001	Telemetry control mode
0010	System control mode
0011	On-board LF oscillator tuning
0100	On-board MF oscillator tuning
0101	On-board HF oscillator tuning
0110	Amplication temperature channel
0111	Voltage reference temp. channel
1000	Calibration activity channel
1001	Sensitivity activity channel
1010	Sampling period
1011	Monitoring period
1100	Voltage reference battery check
1101	Calibration temperature measurement
1110	Switch off microprocessor oscillator
1111	Run with current settings

Table 2.1: Overview of the SIC commands.

- **Monitoring mode:** This mode is controlled by the SIC's finite state machine. When its internal clock generates a monitoring trigger, the SIC wakes up and the temperature, activity and battery status measurements start. To reduce the required on-board data storage capacity and the power consumption (for data transmission), the activity signals (with a bandwidth of 50 Hz) are processed by a circuit on board of the SIC. This is done in such a way that only the mean value of the activity over the measuring period is stored. When all the measurements are finished, the microcontroller is woken up and the captured data are transferred to the RAM of the microcontroller. Hereafter, the SIC and the microcontroller reenter their sleep mode until a new monitoring trigger occurs and the monitoring cycle restarts.

- **Sending data:** The transponder can be read out by activation by an external 132-kHz field. After activation, the microcontroller wakes up and starts to send out the captured data or the identification code of the device. Manchester encoding (cf Section 5.5) is used, resulting in an enhanced communication. The carrier frequency of 66 kHz, which is used for data transmission, is deduced from the 132-kHz activation field. In this way a correct transmission carrier-frequency is obtained, independent of the supply voltage/temperature of the transponder itself, which is beneficial from the technical point of view. From the commercial point of view though, the need for two coils for communication is disadvantageous.

- **Receiving commands:** After data transmission, the device waits for a short period and after this period, it is checked if the 132-kHz field is still present. If this is the case, the device can be (re-)programmed. By modulating (AM) the 132-kHz field, commands can be sent to the device. The decoding of these command is carried out by the microcontroller.

2.4.1.3 Practical realization and problems

The SIC's temperature and activity channel (with on-board compression) have already been presented in [Wou 95]. The movement channel interface is a switched-capacitor circuit which measures the difference between two capacitive accelerometers: a sensitive accelerometer and an identical overdamped accelerometer, sensitive 'only' to low-frequency accelerations. By measuring the difference between the capacitances of the two accelerometers, the DC-component of the movement signal is filtered out. Note that this is an example of analog data processing (cf Section 2.3.3). From the commercial point of view though, the need for two accelerometers is disadvantageous. A new interface, which uses only one accelerometer and also filters out the DC-component (digitally) is introduced in Section 2.4.2.2.

During tests of the prototype, a communication problem between the microcontroller and the SIC showed up. This has been solved by reprogramming the microcontroller and by extending the 8 communication lines (Fig. 2.4) between the SIC and the microprocessor to 9. Due to the unavailability of the original (mask-programmable) microcontroller, a new microcontroller (Philips OTP PCD3745) has been chosen for. This is a One Time Programmable (OTP) microcontroller, giving rise to lower development costs (lower ordering quantities) and a high flexibility towards future software changes. Because of the unavailability of the chip in die form (except for very

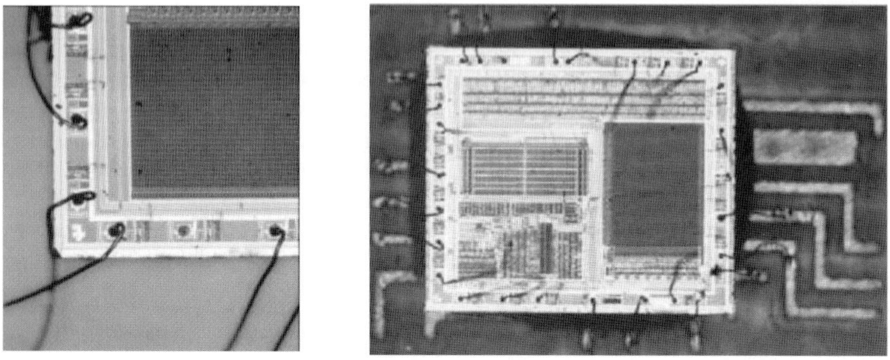

(a) (b)

Figure 2.6: *(a) Decapsulated microcontroller (b) placed on the new substrate.*

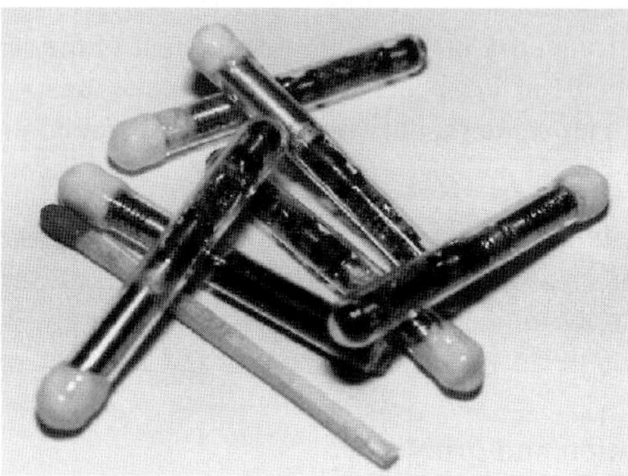

Figure 2.7: *Photograph of a few injectable transponder prototypes.*

large quantities) and the limited project budget, it has been decided to recover the chip out of the package (after programming the code). This has been carried out by etching off the plastic with fumic nitric acid (Fig. 2.6 (a)). Next, the naked microcontroller is placed on a substrate, while the bond wires, still connected to the bond pads of the microcontroller, are glued conductively to the metallic connections on the new substrate (Fig. 2.6 (b)). Tests have proven the functionality of microcontrollers decapsulated with this technique. Though, for reasons of power consumption, data-storage-capacity and area, it is beneficial to use a custom designed FSM instead of a microcontroller, as explained further.

Due to the altered dimensions of the microcontroller and the extension of the number of communication lines, a new substrate for the electronics needed to be developed. To fit all the components in a cylindrical glass capsule of 4 cm by 6 mm (cf Fig. 2.7), the SIC has to be glued non-conductively on top of 4 capacitors, while the naked microcontroller (with conductively glued bond wires) is placed underneath it. It is clear that this is only possible for prototype development and not for commercial devices.

2.4.2 Market introduction: DEVICE 4

2.4.2.1 Modifications

Several technical modifications were necessary to allow a successful commercialization of DEVICE 3. These changes include: making the device compliant with the existing ISO-protocol (for RF IDentification applications), reduction of the number of components (to lower the cost), increase of the on-board data storage capacity, reduction of the power consumption (longer life time) and addition of some new features. Moreover, the application area of the redesigned device, called DEVICE 4, is extended to the food distribution sector yielding a broader market segment.

To get the DEVICE 4 transponder approved for commercial use, it must comply with the existing ISO standards for RFID transponders (11784/5). The ISO 11785 protocol dictates the use of only one frequency band (134 kHz) for communication instead of two (cf DEVICE 3: 132 kHz and 66 kHz). Unfortunately, this protocol is intended for identification purposes only. It does not describe how large amounts of collected sensor data have to be sent from a transponder to an external receiver nor how commands have to be sent to a transponder. To solve this problem an extension of the original ISO protocol has been proposed within the FAME project. The proposed DEVICE 4 protocol is designed in such a way that the transponder will normally act like an ISO-compatible RFID transponder, which can communicate with any off-the-shelf ISO-compatible reader. However, when extra functionality is required, the reader sends a special activation pulse, followed by a command, to the transponder. In this way e.g. the transmission of collected data from the transponder to the reader can be activated. The command/data exchange is checked for correctness by means of a 16-bit CRC-code.

To lower the cost of the transponder the number of components has been reduced. First of all, only one coil is used for communication instead of two. In fact, the total number of external components for the RF-interface has been reduced to two instead of nine (cf DEVICE 3). Another important change is the choice for a one-chip solution instead of the combination of a sensor interface chip and a microcontroller: a FSM with a similar functionality as the microcontroller in DEVICE 3 is now integrated on chip, eliminating the need for an extra component and off-chip interconnections. A last important modification is the change of the accelerometer interface requiring only one accelerometer for the movement measurement instead of two. The total number of external components of the complete transponder has been reduced from 14 (DEVICE 3) to 7 (DEVICE 4).

The DEVICE 4 transponder has a larger memory capacity than DEVICE 3. It includes a total of 1 kbytes of RAM, allowing for 512 samples (activity *and* temperature) to be stored before running out of memory. This allows more frequent measurements over a longer period and results in an extension of the application area of the device.

An important issue during the development of DEVICE 4 has been the reduction of its overall power consumption to extend its life time. Elimination of the microcontroller is one of the major improvements made in this field. A further reduction in current consumption has been

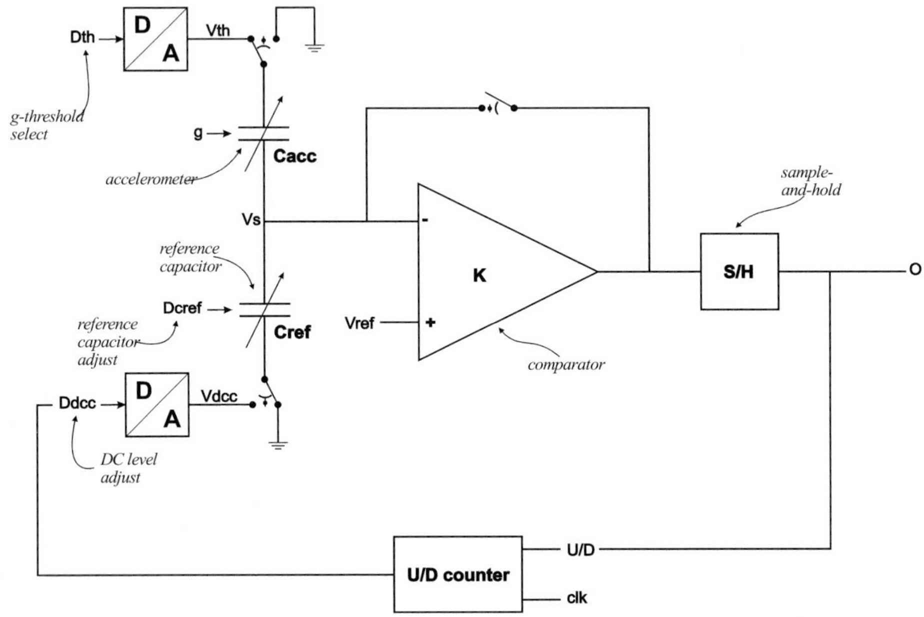

Figure 2.8: *Block diagram of the movement channel.*

made possible by replacing the 3 different oscillators incorporated in DEVICE 3 by 1 and by redesigning the movement-channel sensor-interface (cf further).

Because of the application-area extension to the food distribution sector, DEVICE 4 is equipped with some new features. The temperature channel e.g. supports a new temperature range, especially intended for use in food applications (-10 °C $\leftrightarrow$ +10 °C). Another new feature is the introduction of 'sign off codes'. These sign off codes are used to electronically identify the individual stages in a complete distribution chain. This e.g. allows to trace who is responsible for irregularities during the distribution of a product.

2.4.2.2 Sensor channels

As already mentioned above, the mechanically-overdamped accelerometer is replaced by a digital DC-cancelling circuit in the ***movement channel*** of DEVICE 4. Note that due to the elimination of this accelerometer the power consumption (cf Table 2.2) of the movement channel (Fig. 2.8) could be reduced.

The movement-channel sensor-interface is based on the direct comparison of the capacitance values of the accelerometer C_{acc} and a reference capacitor C_{ref}. After charging one plate of the capacitors to V_{ref} in one phase (unity-follower configuration), the charges on both capacitors are 'subtracted from each other' during the next phase and the resulting voltage V_s (found by the law of charge conservation)

$$V_s = V_{ref} + \frac{C_{acc} \cdot V_{th} - C_{ref} \cdot V_{dcc}}{C_{acc} + C_{ref}} \tag{2.1}$$

is compared with the reference voltage V_{ref}. Note that the comparator is offset-cancelled, based on the CDS technique. The output of the comparator is fed to a sample-and-hold circuit S/H, where it is processed further by digital circuitry: the mean value of the activity over a given period is derived from the digital data stream by means of a counter. Note that the values of V_{th} (sensitivity) and C_{ref} (rest capacity) are programmable in order to be able to interface with different types of accelerometers.

The removal of the overdamped accelerometer in DEVICE 4 implied that a new circuit for the DC-filtering of the acceleration signal had to be developed. Because of the very low cut-off frequency (< 1 Hz) of this filter, an ordinary RC-filter would require too much die area. This is solved by a digital filter, incorporated in the feedback loop (Fig. 2.8) from the sample-and-hold circuit S/H to the voltage V_{dcc} that is applied to the reference capacitor C_{ref}. This filter consists of an up/down counter and a digital-to-analog converter (DAC). The MSB bits of the 13-bit up/down counter track the DC-level of the acceleration signal. After each comparison the value of the counter is adapted (positive or negative). Note that the step taken by the up/down counter after each comparison is programmable. The DAC converts the digital representation of the DC-signal (6 MSB bits of the counter) to the voltage V_{dcc}. This is applied to the reference capacitor C_{ref} for subtraction of the 'DC-signal' from the accelerometer signal (cf Eq. (2.1)).

In contrast with DEVICE 3, the ***temperature channel*** in DEVICE 4 must be capable to measure the temperature in two different ranges: 30 °C $\leftrightarrow$ 40 °C (animal application) and -10 °C $\leftrightarrow$ 10 °C (food application) with an accuracy of 0.1 °C and 0.5 °C respectively. A different type of thermistor is used for both applications.

Fig. 2.9 shows the block diagram of the temperature channel of DEVICE 4. First, a low-voltage reference is created from an internal temperature-independent bandgap voltage reference (A1 and R). Next, the voltage is applied to an external NTC-thermistor (A2 and NTC) and a temperature-dependent current flows through the thermistor. The temperature-dependent current is mirrored and fed to a current-controlled oscillator (CCO). By using the thermistor voltage as the reference voltage for the CCO, any unwanted temperature dependencies in the first part of the circuit are cancelled out automatically.

The output of the CCO, oscillating at a temperature-dependent frequency, is used as a clock signal for a counter (CNTR1), which is the digital representation of the temperature. The only external component in the temperature channel is the NTC-thermistor. Replacing it by an on-chip temperature sensor would reduce the accuracy of the measurement. In that case, the required accuracy of 0.1 °C for the animal application would be hard to meet.

A comparison between the two devices is given in Table 2.2.

2.4.2.3 Practical realization

Fig. 2.10 shows the layout of DEVICE 4 and the indication of the most important building blocks. The chip has been realized in a 2.0-μm CMOS technology (ELMOS). It measures 7.28 mm by 3.05 mm = 22.2 mm^2. Fig. 2.11 depicts the hybrid of DEVICE 4, which is ready to be

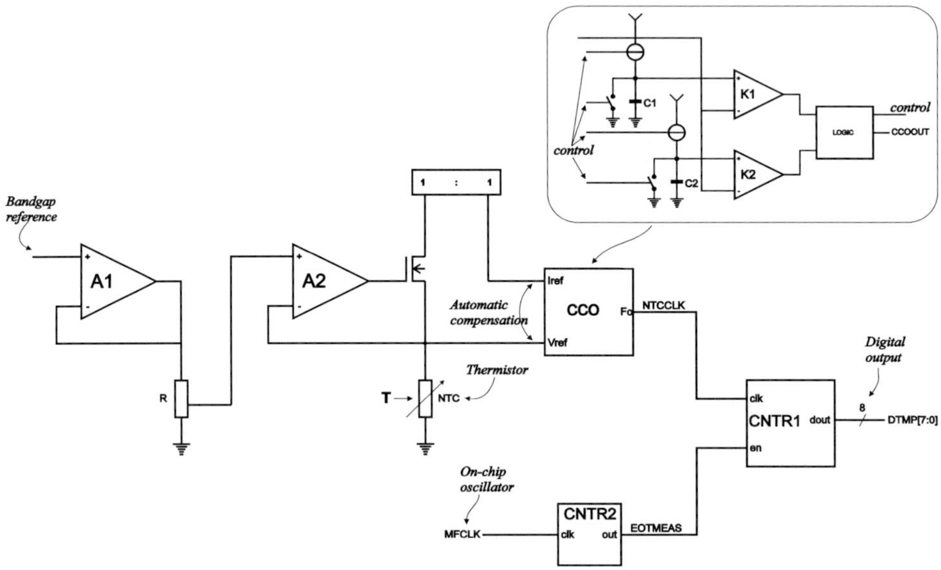

Figure 2.9: *Block diagram of the temperature channel.*

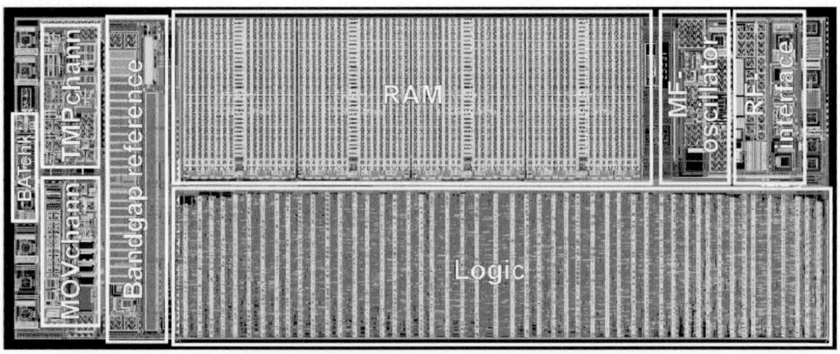

Figure 2.10: *Layout of DEVICE 4.*

Property	DEVICE 3	DEVICE 4
Temperature window	35 °C ↔ 41 °C	-10 °C ↔ 10 °C / 30 °C ↔ 40 °C
Temperature accuracy	0.1 °C	0.5 °C / 0.1°C
Estimated life time	up to 12 months	up to 14 months
Data capacity	128 bytes	1 kbyte
RF activation Þeld	132 kHz	134 kHz
Communication RX		
Carrier frequency	132 kHz	134 kHz
Modulation	ASK	ASK
Encoding	Manchester	Manchester
Bit rate	128 bits/s	500 bits/s
Communication TX		
Carrier frequency	66 kHz	134 kHz
Modulation	ASK/PSK	FSK
Encoding	Manchester	NRZ
Bit rate	172 bits/s	8075 bits/s
Current consumption (@ 3V)		
Temperature conversion	30 μA	33 μA
Activity measurement	12 μA	7.4 μA
Battery check	20μ A	36μţA
LF clock (64 Hz)	115 nA	-
MF clock (8.2 kHz)	5 μA	0.8 μA
HF clock (4.5 MHz)	250 μA	-
Quiescent current	3 μA	2.4 μA

Table 2.2: Comparison between the two devices.

encapsulated in a glass capsule. From left to right, the ferrite coil, the chip, the accelerometer, the thermistor and the battery can be distinguished. To conclude the time scale of the different developments, which have lead to the commercial transponder, is illustrated in Fig. 2.12. The properties of the devices are given in Table 2.3. The transmitters of DEVICE1 and Capt-α are different. Dev 3-MAXI is a larger version of DEVICE3, which has been used to test different activity processing algorithms. This time scale clearly shows that the road from research prototypes to (pre-)commercial devices is long. Many pitfalls (foreseen as well as unforeseen) have to be dealt with on the way.

2.5 Conclusion

In this chapter an overview of the most important design aspects of miniaturized biotelemetry systems has been given. These systems allow to measure physiological parameters without in-

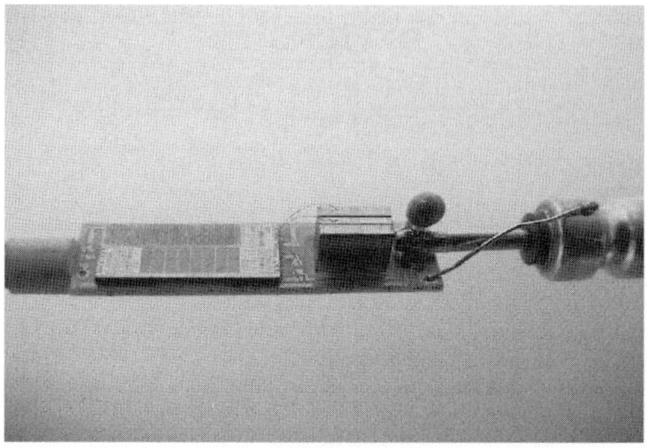

Figure 2.11: *Photograph of the DEVICE-4 hybrid.*

convenience for the subject under surveillance. The most important design issues for implantable biotelemetry devices are: low power consumption, miniaturization, packaging and high reliability. By the implementation of a bi-directional wireless transceiver in the biotelemetry system its flexibility is significantly improved. The bi-directional wireless link allows to reprogram the configuration of the device *in situ* and may offer a solution to cope with long-term drift of sensors.

The general architecture of a telemetric datalogger has been presented. To deal with the different types of non-idealities of real sensors, different strategies, such as calibration, compensation and filtering, can be applied. These functions are implemented in the signal conditioning block, which usually has two other important functions: low-offset and low-noise amplification and analog-to-digital conversion. For the amplifier two different topologies, both including offset and 1/f-noise cancellation, can be selected: the chopper amplifier and the switched-capacitor amplifier based on the CDS technique. The former is more appropriate for continuous-time applications, whereas the switched-capacitor amplifier is more suited for sampled-data applications, where aliasing is unavoidable. The next block is the data processing unit, which extracts rele-

Device	**Continuous**	**Transponder**	**Temperature**	**Activity**
Telemetric Temperature Unit	**X**		**X**	
DEVICE1	**X**		**X**	**X**
DEVICE2		**X**	**X**	
Capt-	**X**		**X**	**X**
Dev 3-MAXI		**X**	**X**	**X**
DEVICE3		**X**	**X**	**X**
DEVICE4		**X**	**X**	**X**

Table 2.3: Device properties.

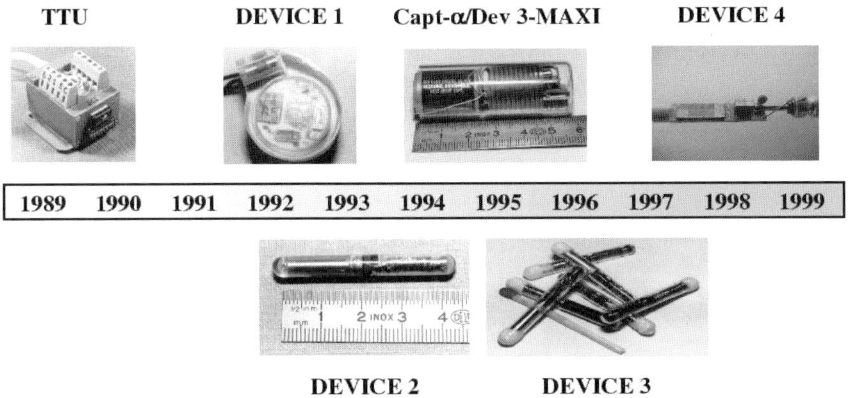

Figure 2.12: *Time scale of the developments.*

vant information from the sensor signals and hence reduces the required data storage capacity and indirectly enhances the efficiency of the communication link. Digital data processing can be implemented by a microcontroller or a FSM. A microcontroller provides flexibility towards future changes, while a FSM yields a lower power consumption and a smaller size. Biotelemetry systems can be divided into two main categories with respect to their powering: active devices and passive devices. The former allow continuous monitoring and have a larger communication range, while the latter give a prolonged life time and possibly a reduction in volume and weight. Passive devices derive their power from an external RF-powering field. The demand for the proximity of this field is an important drawback. For the wireless communication link usually RF techniques/carriers are employed. Low-frequency carriers offer short-range communication, and can propagate through tissue, which makes them convenient for implanted systems. High-frequency carriers on the other hand provide a faster communication and a larger communication distance, but due to their 'inability' to pass through tissue they are not suited for implantable systems. An often underestimated task is the packaging of a datalogger, which must protect the sensors and the electronics from the environment while it must be transparent for the measured parameter. Sensors, which can be hermetically sealed within the package, are most convenient. The concept of smart sensors also has been introduced. Smart sensors combine sensors and sensor electronics on the same chip, offering a lower cost with an improved reliability. On the other hand, co-integrated sensors in a smart sensor system are often less performant due to the conflicting demands of sensors and sensor electronics with respect to their fabrication process. Moreover, the sensor electronics may be degraded by the environmental conditions to which the smart sensor chip must be exposed.

To conclude two examples of implantable dataloggers have been given. The first device is an injectable transponder, used for animal identification and quantification of animal welfare. The second device is a redesigned version of the previous one, intended for commercial use. The encountered problems and the modifications required for a successful commercialization of the injectable datalogger were summarized. It is clear that the road from prototype development to market introduction is a long one with many pitfalls.

Chapter 3

Miniaturized datalogger for stress monitoring in oral implants

3.1 Introduction

After the general introduction on biotelemetry systems given in the previous chapter, a dedicated miniaturized low-power datalogger for *in vivo* stress monitoring in oral implants is introduced in this chapter. More in particular, the system concept of this device is presented here. The following chapters describe its realization.

In this chapter a discussion about the clinical background of the presented work is given and the motivation to develop the miniaturized datalogger is explained. Dental prostheses are kept in place by oral implants. Unfavorable loading conditions are generally accepted as the most important cause of implant failure. Due to the lack of quantitative *in vivo* load data, it is necessary to measure the loads acting on the implants to gain more insight. The employed measurement methodology is discussed. Strain gauges are installed on the abutments, i.e. (hollow) cylinders in the gums on top of the implants. By measuring the strain gauge resistance values the loads on the different abutments can be derived. A first important step in the measurement of *in vivo* loads on oral implants/abutments has been the development of an external non-portable measurement system. Due to the drawbacks associated with this external system, a new miniaturized low-power datalogger, embedded within the dental prosthesis itself, is required. A comparison between silicon strain gauges and metal film strain gauges is made and self temperature compensated strain gauges are introduced. To conclude the specifications of the new miniaturized datalogger are described.

3.2 Clinical background and motivation

The principle of osseointegration has been discovered by the Swedish Professor P.-I. Brånemark in the 1960s. He found that pure titanium implanted in the bone integrates with the bone without an intervening layer of fibrous soft tissue. The genetic code that commonly makes bone reject a foreign material is not activated when titanium is employed as implant material. Instead, nature allows the direct attachment of bone cells to the titanium surface (see Fig. 3.1), giving rise to a rigid connection between the titanium and the bone. This allows to employ titanium implants as anchorages for prosthetic reconstructions. From a biomechanical point of view, an implant is considered osseointegrated if there is no progressive relative motion between the implant and the surrounding bone under functional levels and types of loading for the entire life of the patient and the implant exhibits deformations of the same order of magnitude as when the loads would be applied directly to the bone [Duy 00].

Based on the osseointegration principle titanium implants have been employed since 1965 to treat partially and fully edentulous patients. A schematic drawing of an oral implant of the Brånemark type together with a dental prosthesis is shown in Fig. 3.2. This type of implant is one example of the large variety of implant systems existing today on the market. The complete Brånemark system consists of: 1) an implant, 2) an abutment screw (fixating the abutment), 3) an abutment (cylinder between the implant and the oral cavity through the gums (i.e. mucosa)), 4) a set screw (fixating the prosthesis), 5) a gold cylinder (to be casted in the prosthetic superstructure) and 6) artificial tooth, called prosthetic superstructure. The latter can be a partial prosthesis, replacing a single tooth (as shown in Fig. 3.2) or a number of teeth, or can be a full prosthesis, replacing all the teeth of the lower jaw (mandible) or the upper jaw (maxilla). At first stage surgery, the placement of the titanium implants is done by progressively drilling wider cavities into the bone and screwing the implants into the resulting holes. The implants are then allowed to heal submucosally without direct-load application in order to prevent relative motion at the implant-bone interface. At second stage surgery, after an healing period of 3 to 6 months, depending on the quality of the bone where the implants are located, the abutments are placed upon the implants and fixed with abutment screws (cf Fig. 3.2). Two major classes of full prostheses can be distinguished: implant-supported fixed prostheses and implant-retained removable overdentures. Implant-supported fixed prostheses contain a horse-shoe shaped (metal) framework with artificial teeth on top and with fitting holes to attach to the implants by set screws (cf Fig. 3.2). The prosthesis is completely fixed by the implants and the patient is not able to remove this kind of prosthesis. Fig. 3.3 shows a mandible equipped with five abutments, fixed onto the implants and a fixed prosthesis placed on these implant-abutment entities. In contrast to implant-supported fixed prostheses, implant-retained overdentures are attached to the implants in a way that permits the patient to remove them at any time. This kind of prosthesis is not only supported by the implants, but by the implants *and* the patient's mucosa. The prosthesis can be retained by two different attachment types, as shown in Fig. 3.4: ball-type abutments and bar attachments, both allowing to snap prostheses with matching cavities onto them. In the case of implant-retained prostheses, the main function of the implants is to offer more comfort to the patient in comparison with classic non-retained conventional dentures.

Despite the optimistic success rate achieved with oral implants, failures do occur and are believed to be the result of two major causes or a combination of both [Duy 00, Oos 00]: mechanical implant failure and biological implant failure. An implant is considered successful when parameters such as mechanical function (ability to speak, chew), tissue physiology (presence of osseointegration, maintenance of supporting bone, absence of inflammation), and psychology (absence of pain and discomfort, aesthetics) reach an acceptable level [Mom 94]. While a mechanical failure refers to a fracture of an implant component, mostly due fatigue, a biological

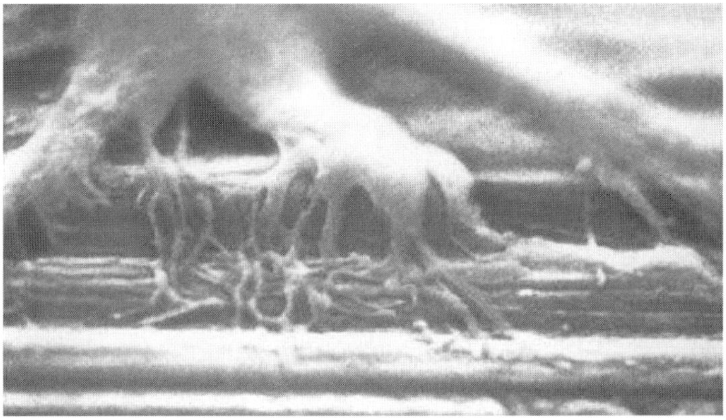

Figure 3.1: *Scanning electron micrograph showing bone tissue attaching to titanium (supplied by P.-I. Brånemark).*

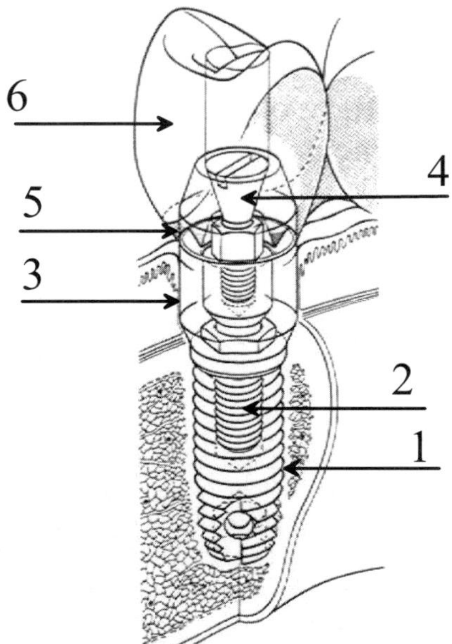

Figure 3.2: *Schematic drawing of a Brånemark implant together with a tooth replacement. 1) implant, 2) abutment screw, 3) abutment, 4) set screw, 5) gold cylinder, 6) artificial tooth.*

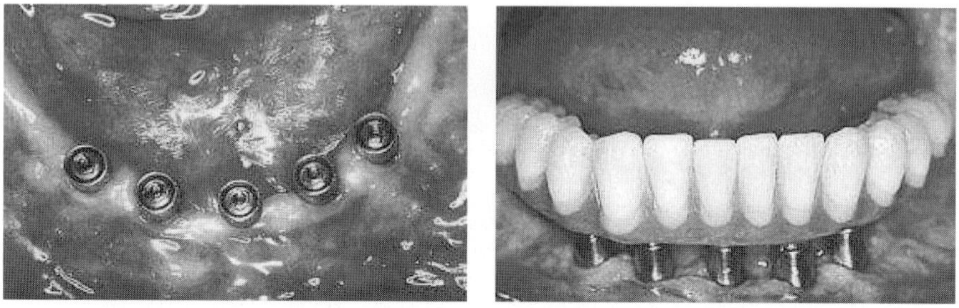

(a) (b)

Figure 3.3: *(a) Mandible equipped with 5 abutments fixed onto the implants. (b) Implant-supported fixed prosthesis placed on these abutments.*

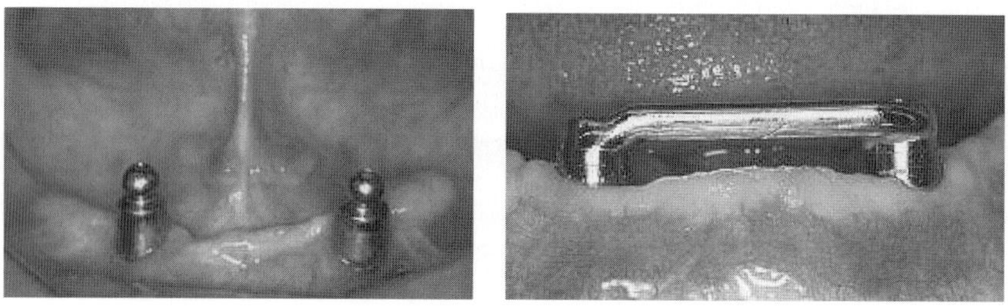

(a) (b)

Figure 3.4: *Two different attachment types for implant-retained overdentures: (a) ball-type abutments and (b) a bar attachment.*

implant failure is caused by the inadequacy to establish or maintain osseointegration. A distinction must be made here between early implant failures and late implant failures [Esp 98a]. Early implant failures occur in the first weeks or months after the implant installation, during the healing phase, and result from the inability to establish osseointegration, whereas the inability to maintain the achieved osseointegration for stable implants, under functional conditions, gives rise to late failures. Biological implant failure has different faces, but mainly implies the loss of osseointegration or continuous marginal bone loss. When there is a loss in direct contact between the titanium implant and the bone, mostly characterized by mobility of the implant, a fibrous tissue develops at the interface between the implant and the bone. Because of the weak nature of this fibrous encapsulation of the implant in comparison with osseous integration of the implant, the fixation is not stable anymore, causing implant failure. Biological failure is also associated with excessive marginal bone loss, which is the loss of bone around the neck and shoulder of the implant. Since the conservation of this bone is crucial for long-term implant success, it is often used as a prognostic criterion [Mom 94].

Late biological implant failures are thought to be caused by two major factors: infection and mechanical loading. Inflammatory processes can disrupt the biological equilibrium of the soft tissues around the implant-abutment entities, possibly giving rise to marginal bone loss. However, more long-term studies are necessary to confirm this theory of infection-induced implant failures [Duy 00, Oos 00]. Based on *in vitro*, animal, and human studies, it has been suggested that most biological implant failures are caused by a disequilibrium of the loads acting on oral implants [Esp 98b]. Overloading as well as underloading, related to the quality and the quantity of the bone surrounding the implants, are put forward to be possible causes of marginal bone loss. It has been recognized that marginal bone loss can be correlated with overload originating from unfavorable prosthesis design and parafunctional habits (clenching, bruxism). Also the risk of excessive marginal bone loss due to high bending moments has been confirmed by many authors. In contrast with the overload theory, others have suggested underloading to be responsible for marginal bone loss. Bone has the property to adapt its geometry and its internal structure to resist to imposed forces, a phenomenon which is called bone (re)modeling. Due to this property, the amount of bone can increase on heavy loaded spots and can also decrease, giving rise to implant failure, on underloaded spots (i.e. disuse atrophy). The biomechanical aspects of the processes, involved in the (re)modeling of the bone around loaded oral implants, remain uncertain and require more investigation [Hos 94].

Several clinical studies have put forward unfavorable loading conditions as the most important factor giving rise to implant failures. In none of these studies, however, an attempt was made to quantify the *in vivo* implant loads or the *in vivo* bone loading and to relate these to the observed marginal bone loss and/or implant failures [Oos 00]. In order to verify the load-related failure theories presented in these studies and to gain more insight in the biomechanical processes involved in bone (re)modeling and implant failures, there is a strong need for quantitative data, both at the implant level and at the bone level. Since true clinical conditions still can not be mimicked accurately in theoretical models or even in *in vitro* experiments [Duy 00], it is necessary to acquire quantitative data from *in vivo* measurements. To perform *in vivo* measurements of the loads imposed on the implants during (para)functional loading (chewing, maximal biting, clenching, ...) of a dental prosthesis and to relate the measured load data to clinically observed bone (re)modeling around the implants for individual patients, a multidisciplinary research team, consisting of the ESAT-MICAS group, the Division of Biomechanics and Engineering Design and the Department of Prosthetic Dentistry of the K.U.Leuven, has been formed. The role of the ESAT-MICAS group within this research project has been the development of a measurement system which is capable of registrating the *in vivo* loads, imposed on the different implant-abutment entities (sometimes also called 'implants' in this work), supporting/fixating the prosthesis. The development of patient-dependent Finite Element models, using the measured implant loads as input parameters, has been performed by the Division of Biomechanics and Engineering Design. These models allow to translate the measured *in vivo* implant-abutment loads to *in vivo* bone tissue loads by employing Finite Element analysis. The patient-dependent models are based on the bone anatomy of the patient, the bone's elastic properties, the implants' placement, the prosthesis design, and the measured implant loads. To create patient-dependent Finite Element models, X-ray computed tomography (CT) images of the patient's jaw are used. The Finite Element models and the results obtained with these models are presented in detail in [Oos 00]. The measurements of the loads imposed on the implants for several types of dental prostheses as well as animal experimental studies have been performed by the Department of

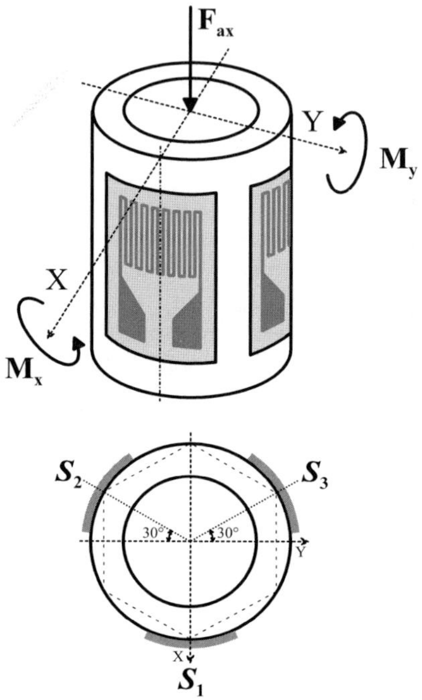

Figure 3.5: *Schematic representation (top) and top view (bottom) of an abutment equipped with 3 strain gauges (gray).*

Prosthetic Dentistry. The results of these load measurements and animal studies are described extensively in [Duy 00]. The ultimate goal of the presented project is to gain more insight in the bone (re)modeling processes in order to stimulate preventive actions and further increase the predictability of the treatment outcome obtained with osseointegrated oral implants.

3.3 Measurement methodology

In order to measure the *in vivo* loads acting on the different titanium implant-abutment entities supporting a dental prosthesis, strain gauges are installed on the abutments (cf Fig. 3.2). An abutment of the Brånemark implant system (Nobel Biocare®, Gothenburg, Sweden), applied during the experiments, can be represented as an hollow cylinder with an outer diameter of 4.5 mm and an inner diameter of 3 mm, with height values ranging between 5.5 mm and 7 mm to accomodate matching tolerances with respect to the prosthesis. To measure the loads imposed upon a Brånemark abutment, it is equipped with 3 strain gauges, placed 120° from each other, with their measuring grids parallel to the cylinder axis, as shown in Fig. 3.5. The strain gauges are

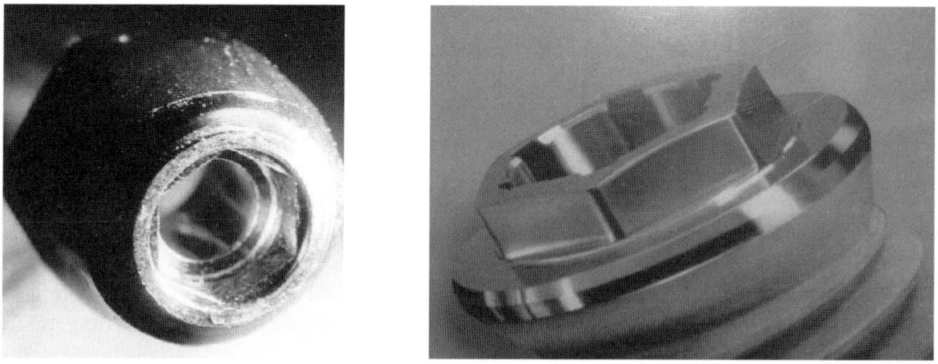

(a) (b)

Figure 3.6: *(a) Internal hexagon located at the base of a Brånemark abutment. (b) Matching hexagon located on top of an implant.*

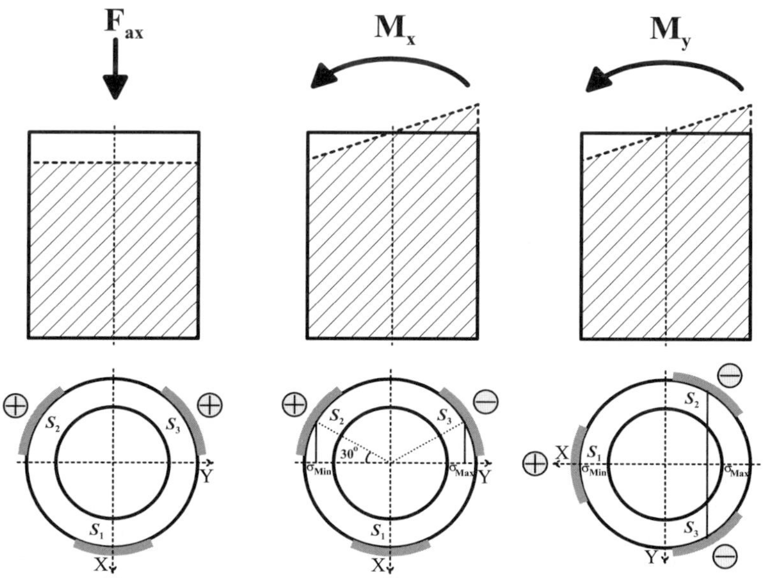

Figure 3.7: *Overview of the measured load components and the induced deformations (shaded) of the cylinder (side view and top view) by these load components.*

centered on the corners of the internal hexagon at the base of an abutment, which is depicted in Fig. 3.6(a). This internal hexagon is used to position the abutments on the implants, which have a matching external hexagon located at their top (see Fig. 3.6(b)). The definition of the names of the strain gauges, the coordinate system, and the measured load components, is also illustrated in Fig. 3.5. The position of the coordinate system, linked with the abutment, is determined by the abutment's position, which in turn is determined by the implant's position, since the two are connected by matching hexagons. When an abutment is placed upon an implant, it is positioned in such a way that a strain gauge is located on the hexagon's corner, closest to the buccal side (opposite to the tongue's side) of the jaw's arc. By definition this strain gauge is named S_1 and the other strain gauges, clockwise, S_2 and S_3. The X-axis is defined as the cylinder's top cross-section central line, which is perpendicular to the central line of S_1. Three different load components are measured: the axial force F_{ax} along the axis of the cylinder, the bending moment M_x around the X-axis, and the bending moment M_y around the Y-axis. The positive direction of these bending moments is indicated by the arrows near M_x and M_y and the axial force F_{ax} is defined positive if the cylinder is compressed.

Fig. 3.7 gives an overview of the deformations of the hollow cylinder caused by the individual measured load components F_{ax}, M_x and M_y. Due to the cylinder's deformation the installed strain gauges are also transformed. Shortening (i.e. compression) of the strain gauges is indicated with a positive sign and elongation (i.e. tension) with a negative sign. The transformation of the strain gauges causes a change in their resistance value and by measuring the strain gauge resistance values and combining them in the right way, the load components of an arbitrary load imposed upon the abutment can be reconstructed. The individual measured load components give rise to specific transformation patterns of the strain gauges:

- A positive axial force component F_{ax} gives rise to an equal transformation (i.e. compression) of the three strain gauges.

- A positive bending moment M_x around the X-axis results in compression of S_2 and tension of S_3. The resistance changes of S_2 and S_3 are equal in size, because they are located at the same distance from the X-axis, but opposite in sign, since S_2 is under compression and S_3 under tension. Half of S_1 is under compression, while the other half is under tension, resulting in an unaffected resistance value for S_1.

- A positive bending moment M_y around the Y-axis results in compression of S_1 and tension of S_2 and S_3. Since S_2 and S_3 are both under tension and located at the same distance from the Y-axis, their resistance changes are equal in size and equal in sign. The sum of the resistance changes of S_2 and S_3 is equal in size and opposite in sign to the resistance change of S_1, because their distance to the Y-axis is half of the distance of S_1 to the Y-axis, and S_2 and S_3 are under tension, while S_1 is under compression (first-order approximation).

Based on the deformations and the resulting resistance changes of the installed strain gauges, caused by the individual load components, the equations to calculate the load components F_{ax}, M_x and M_y of an arbitrary load, imposed upon the abutment, can be derived. The relation between the axial force load component F_{ax} and the strain gauge resistance values is given by

Eq. (3.1), where $\mathrm{R}(S_\mathrm{i})$ denotes the resistance value of S_i and R_0 the resistance value of the strain gauges when no load is imposed upon the abutment. This equation results from the observation that the contribution to the mean value of the strain gauge resistance changes by $\mathrm{M_x}$ and $\mathrm{M_y}$ load components is zero, since the resistance changes of S_1, S_2 and S_3 due to a $\mathrm{M_x}$ load component and also due to a $\mathrm{M_y}$ load component cancel out each other. This means that the mean value of the strain gauge resistance changes is proportional to the axial force load component of the imposed load. The negative sign is added to $\mathrm{F_{ax}}$, because a positive axial force load component gives rise to negative resistance changes of the strain gauges (under compression).

$$-\mathrm{F_{ax}} \sim \Delta\mathrm{R_{F_{ax}}} = \frac{[\mathrm{R}(S_1)-\mathrm{R}_0]+[\mathrm{R}(S_2)-\mathrm{R}_0]+[\mathrm{R}(S_3)-\mathrm{R}_0]}{3} \tag{3.1}$$

$$= \frac{\mathrm{R}(S_1)+\mathrm{R}(S_2)+\mathrm{R}(S_3)}{3} - \mathrm{R}_0 \tag{3.2}$$

The bending moment load component $\mathrm{M_x}$ around the X-axis is related to the strain gauge resistance values through Eq. (3.3). This equation results from the fact that the resistance changes of S_2 and S_3 due to $\mathrm{F_{ax}}$ and $\mathrm{M_y}$ load components are equal in size and equal in sign, whereas a $\mathrm{M_x}$ load component gives rise to resistance changes in S_2 and S_3 equal in size and opposite in sign. The difference in resistance between S_2 and S_3 is thus independent of $\mathrm{F_{ax}}$ and $\mathrm{M_y}$ load components and proportional to the $\mathrm{M_x}$ load component of the imposed load. The factor $\frac{1}{2}$ is introduced to obtain the resistance change of S_2 only and the negative sign is added to $\mathrm{M_x}$, because a positive bending moment component $\mathrm{M_x}$ around the X-axis gives rise to a negative resistance change of S_2.

$$-\mathrm{M_x} \sim \Delta\mathrm{R_{M_x}} = \frac{\mathrm{R}(S_2)-\mathrm{R}(S_3)}{2} \tag{3.3}$$

The bending moment load component $\mathrm{M_y}$ around the Y-axis is related to the strain gauge resistance values through Eq. (3.4). The resistance of S_1 does not change due to a $\mathrm{M_x}$ load component, but does change due to both the $\mathrm{F_{ax}}$ and $\mathrm{M_y}$ load components of the imposed load. The difference between the total resistance change of S_1 and the resistance change due to the $\mathrm{F_{ax}}$ load component (i.e. $\Delta\mathrm{R_{F_{ax}}}$) is proportional to the $\mathrm{M_y}$ load component.

$$-\mathrm{M_y} \sim \Delta\mathrm{R_{M_y}} = [\mathrm{R}(S_1)-\mathrm{R}_0] - \Delta\mathrm{R_{F_{ax}}} = \mathrm{R}(S_1) - \frac{\mathrm{R}(S_1)+\mathrm{R}(S_2)+\mathrm{R}(S_3)}{3} \tag{3.4}$$

The magnitude of the overall bending moment M, imposed upon the abutment, is found by the vector summation of the bending moment load components around both axes $\mathrm{M_x}$ and $\mathrm{M_y}$ and is given by Eq. (3.5)

$$\mathrm{M} = \sqrt{\mathrm{M_x}^2 + \mathrm{M_y}^2} \tag{3.5}$$

To find the relation between the resistance changes of the strain gauges and the imposed load components, first the relation between the relative deformation $\varepsilon = \frac{dl}{l}$ (strain) of the titanium hollow cylinder and the imposed load components is calculated. An axial force load component $\mathrm{F_{ax}}$ causes a stress σ in the hollow cylinder, equal to Eq. (3.7), with r_{out} and r_{in} respectively the outer and inner radius of the hollow cylinder's cross-section [Roa 75]. Note that the sign of F_{ax} is not taken into account in this equation.

$$\sigma = \frac{F_{ax}}{A} \tag{3.6}$$

$$= \frac{F_{ax}}{\pi \cdot (r_{out}^2 - r_{in}^2)} \tag{3.7}$$

The relation between the induced stress σ and the relative (elastic) deformation ε of the cylinder is given by Hooke's law (Eq. (3.8)) with $E_{titanium} = 110000 \frac{N}{mm^2}$ the modulus of Young for titanium and l the hollow cylinder's height.

$$\varepsilon = \frac{dl}{l} = \frac{\sigma}{E_{titanium}} \tag{3.8}$$

Combining Eq. (3.7) and Eq. (3.8) gives the relation between an axial force load component F_{ax} and the induced strain ε:

$$\frac{F_{ax}}{\varepsilon} = \pi \cdot (r_{out}^2 - r_{in}^2) \cdot E_{titanium} = 0.97 \frac{N}{\mu strain} \tag{3.9}$$

A bending moment load component $M_{x,y}$ around the X-axis/Y-axis results in a maximum stress, equal to

$$\sigma_{Max} = \frac{M_{x,y} \cdot r_{out}}{I} \tag{3.10}$$

with I, the second moment of inertia of the hollow cylinder's cross-section, given by [Pol 94]

$$I = \frac{\pi}{64} \cdot (d_{out}^4 - d_{in}^4) \tag{3.11}$$

with d_{out} and d_{in} respectively the outer and inner diameter of the hollow cylinder's cross-section. For a bending moment M_x around the X-axis the maximum stress σ_{Max} occurs on the crossings of the outer border of the cylinder's top cross-section with the Y-axis (see Fig. 3.7) and for a bending moment M_y around the Y-axis on the crossings with the X-axis. The maximum stress is constant over the entire height of the cylinder and is negative (i.e. compressive stress) on one side of the cylinder and positive (i.e. tensile stress) on the other side. Combination of Eq. (3.10), Eq. (3.11) and Eq. (3.8) yields the relation between a bending moment load component $M_{x,y}$ and the maximum induced strain ε_{Max}:

$$\frac{M_{x,y}}{\varepsilon_{Max}} = \frac{\pi}{64} \cdot \frac{(d_{out}^4 - d_{in}^4) \cdot E_{titanium}}{r_{out}} = 0.079 \frac{N.cm}{\mu strain} \tag{3.12}$$

The relation between the relative deformation ε and the induced resistance change for a strain gauge is given by Eq. (3.13):

$$\frac{\Delta R}{R_0} = G \cdot \frac{dl}{l} = G \cdot \varepsilon \tag{3.13}$$

where G is the gauge factor and l the length of the strain gauge. Since the strain gauges are glued to the abutment, they are exposed to the same relative deformation ε as the abutment, when a load is imposed on it. Combination of Eq. (3.9), Eq. (3.12), and Eq. (3.13) yields the relations between the strain gauge resistance changes and the imposed load components F_{ax}, M_x and M_y, summarized in Table 3.1. The two rows of Table 3.1 correspond with the two types of strain gauges used during the measurements, as explained further in Section 3.5. Note the difference between the ratios for the M_x and M_y load components. This results from the fact that the maximum stress due to a M_x load component occurs on the crossings of the outer border of the cylinder's top cross-section with the Y-axis, while the strain gauges S_2 and S_3, sensitive to the M_x load component, are located at 30° from this axis (Fig. 3.7). Therefore, a sensitivity factor must be applied to account for the smaller stresses, occurring at S_2 and S_3. The stress due to a M_x load component in a point of the cylinder's top cross-section, located at a distance d from the X-axis is given by

$$\sigma(d) = \frac{M_x \cdot d}{I} \tag{3.14}$$

and this stress is constant over the entire height of the cylinder. From Eq. (3.14) the sensitivity factor can be derived, which equals $\sqrt{3}/2$, i.e. the ratio of the distance between S_2 and the X-axis ($\sqrt{3}/2 \cdot r_{out}$) to the distance between the point(s) of maximum stress and the X-axis (r_{out}). To obtain the same proportionality-factor for M_x and M_y (cf Table 3.1) the strain gauge resistance values in Eq. (3.3) must be multiplied by $2/\sqrt{3}$. The resulting relation between M_x and the strain gauge resistance values equals

$$-M_x \sim \Delta R_{M_{x'}} = \frac{R(S_2) - R(S_3)}{\sqrt{3}} \tag{3.15}$$

In the derivation conducted so far the placement of the strain gauges is assumed ideal. In reality the positioning of the strain gauges is not completely perfect, since they are applied manually on the abutment. Also the nominal resistance value R_0 and the gauge factor G of the different strain gauges are assumed to be the same for each individual strain gauge, while in reality the nominal resistance value and the gauge factor of the strain gauges have a certain tolerance. A last simplification made in the analysis is the absence of an abutment screw. To account for the non-ideal placement of the strain gauges, the tolerance of R_0 and G, and the presence of an abutment screw, a calibration setup together with dedicated calibration software has been developed. To deal with the non-idealities, a set of calibration factors is derived for each abutment before the

Type	**G**	**R_0**	**$R_{F_{ax}}/F_{ax}$**	**R_{M_x}/M_x**	**$R_{M_{y,x}}/M_{y,x}$**
Low-Ω	2.05	120 Ω	-0.25 mΩ/ N	-2.69 mΩ/ N.cm	-3.11 mΩ/ N.cm
High-Ω	2.01	5000 Ω	-10.36 mΩ/ N	-110.17 mΩ/ N.cm	-127.21 mΩ/ N.cm

Table 3.1: Relation between the resistance changes and the imposed loads for two types of employed strain gauges.

Figure 3.8: *Brånemark abutment equipped with 3 strain gauges.*

actual measurements start. These calibration data are then used to obtain the corrected load components from the measured strain gauge resistance values. The calibration setup and software are both described in detail in [Lie 99].

Fig. 3.8 shows a photograph of a Brånemark abutment, equipped with 3 strain gauges. The strain gauges are glued on the abutment, parallel with the abutment's axis, spaced 120° from each other, centered on the corners of the internal hexagon at the abutment's base. A dedicated mounting setup has been developed to enable repeatable and more accurate placement of the strain gauges. To reduce the risk of damaging the strain gauges, bondable printed-circuit terminals, glued between the strain gauges on the abutments, are employed to connect the delicate wires from the strain gauge solder tabs (cf Fig. 3.13) and the wires from the measurement system. The purpose of these terminals is to provide an anchor for both sets of wires, and to prevent forces, transmitted along the measurement system wires, from damaging (e.g. lifted or dislodged solder tabs) the strain gauges or degrading their performance. Moreover, a stress relief loop is introduced in the jumper wires between the strain gauge tabs and the terminals in order to minimize the forces applied to the tabs, and to prevent wire failures at the solder joints. After glueing the strain gauges and the terminals on the abutment and connecting the different wires, a layer of silicones, a shrink sleeve and another layer of silicones are applied to the instrumented abutment to insulate the strain gauges and the connections from the wet oral environment in order to avoid short circuits. The wires connecting the terminals and the measurement system are interwoven to reduce the influence of possible interfering electromagnetic fields. Fig. 3.9 shows two practical examples where the original abutments have been replaced by instrumented ones. On the left 6 implants with instrumented abutments are shown, and on the right a fixed full prosthesis, supported by implants with instrumented abutments, is shown.

Note that in this work the loads at the implant level are derived by means of Finite Element Analysis using the measured loads on the abutment as input parameters and that no direct measurements at the implant level can be performed. Apart from the fact that force sensors with

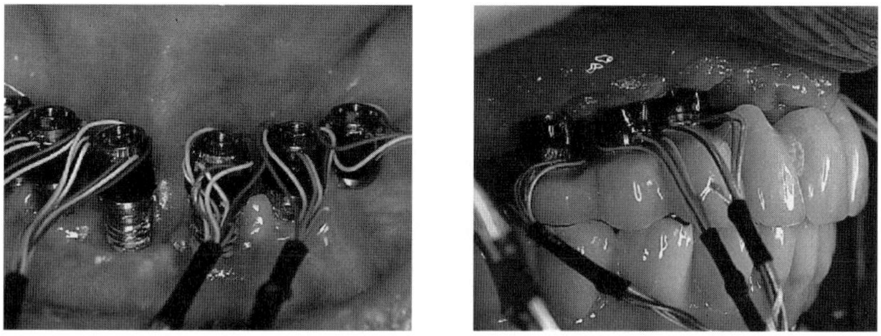

(a) (b)

Figure 3.9: *(a) Implants with instrumented abutments. (b) Fixed full prosthesis supported by implants with instrumented abutments in the upper jaw.*

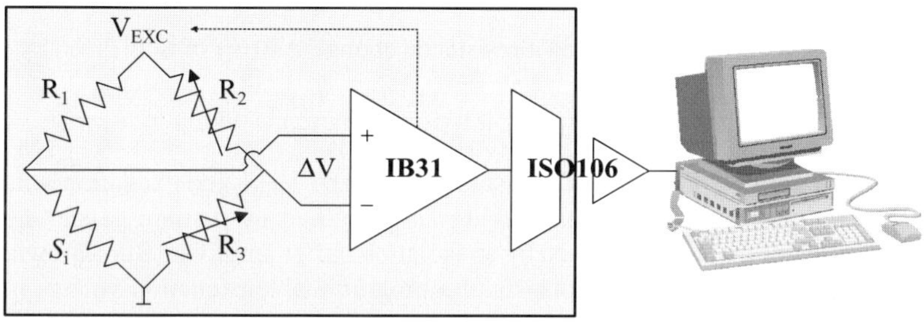

Figure 3.10: *Overview of a single signal path of the external measurement system.*

appropriate dimensions to fit into the implants are not available at present time, there is a benefit in this approach; the patient does not have to undergo an extra surgical treatment for the installation of the measurement system. The original abutments can be replaced by instrumented ones and after the measurement period the patient's original abutments are restored without further implications for the patient.

3.4 External measurement system

A first important step in the qualification and quantification of *in vivo* loads on oral implants has been the development of an external non-portable measurement system. This system, described in detail in [Lie 99], is capable of measuring simultaneously up to 18 different strain-gauge channels. Every strain gauge S_i is placed in a separate Wheatstone bridge (Fig. 3.10) and each bridge has two adjustable resistors R_2 and R_3 in order to balance the bridge. The relation between

Figure 3.11: *External measurement system and connector box.*

the output voltage ΔV of the bridge and a resistance change $\Delta R(S_i)$ of S_i is given by [Lie 99]

$$\Delta V = \frac{\Delta R(S_i) \cdot V_{EXC}}{2 \cdot (2 \cdot R_0 + \Delta R(S_i))} \approx \frac{\Delta R(S_i)}{R_0} \cdot \frac{V_{EXC}}{4} = G \cdot \varepsilon \cdot \frac{V_{EXC}}{4} \tag{3.16}$$

if R_1=R_2=R_3=R_0 with R_0 the nominal resistance of S_i and V_{EXC} the excitation voltage of the bridge. The output voltage ΔV of each bridge is amplified by a strain gauge signal conditioner (Analog Devices, 1B31), followed by an isolation buffer amplifier (Burr-Brown, ISO106). The strain gauge signal conditioner consists of a programmable excitation-voltage block and a programmable-gain instrumentation amplifier, followed by an adjustable two-pole low pass filter. The isolation buffer amplifier is used to isolate the patient from the electrical mains to ensure a safe operation of the measurement system. The amplified analog signals of the 18 channels are digitized by a PC data acquisition card (Microstar Labatories, DAP-800/102) and further processed by a PC. A dedicated Visual Basic computer program, described in [Lie 99], has been developed for calibration, processing and visualization of the collected digital data. Fig. 3.11 shows a photograph of the measurement system with in front of it a connector box, used to connect the wires from the different strain gauges, which are assembled into a separate connector for each abutment. The measurement system also has an additional connection to receive information from a bite fork.

The external system has been employed in the hospital to perform *in vivo* load measurements to investigate the influence of different prosthesis parameters on the occurring loads. The difference in the occurring *in vivo* loads for different prosthesis types has been studied. Also the influence of the number of supporting implants, the prosthesis material, and the attachment system (for overdentures) has been investigated. The results of all these studies can be found in [Duy 00]. After installation of the prosthesis with instrumented abutments, and balancing manually the different strain-gauge channels, the patients are instructed to carry out dental activities during which the loads on the different abutments are measured. Fig. 3.12 shows for instance the measured bending moment on one of the abutments, while the patient is instructed to chew some bread.

Although this measurement system is considered to be a major step forward in the measurement of *in vivo* loads on oral implants, it has several drawbacks:

- The measurement system is external, which implies that the wires from the strain gauges to the measurement system have to come out of the patient's mouth, as shown in Fig. 3.9. Since these wires disturb the normal chewing behavior of the patient, artificial chewing behavior is introduced in the measurements.
- In addition the measurements are restricted to the hospital environment and are done on command, which also introduces artificial chewing behavior in the measurements.
- Another drawback is that the way of measuring does not allow to measure unconscious nocturnal dental activities like bruxism and clenching, which are seen as a missing link for the validation of existing bone remodeling models.
- A last disadvantage of the system is that the balancing of the 18 strain-gauge channels must be done manually.

These drawbacks clearly indicate the need for a miniaturized measurement system, part of the prosthesis and capable of measuring continuously over a longer period. In this way the measurements can be carried out in the normal living conditions of the patient, independent of the hospital environment, so that artificial chewing behavior is kept to a minimum. With this miniaturized system the patient is not bothered any longer with wires coming out of his/her mouth and in addition the measurements can go on during the night without any further inconvenience for the patient so that valuable information about unconscious nocturnal dental activities (i.e. bruxism and clenching) can be collected. Moreover, the ease of use of the miniaturized datalogger is improved by extending the functionality of the datalogger with an automatic compensation block, which can be activated by a bi-directional wireless link, as explained further.

3.5 Strain gauges

Before the new miniaturized datalogger is discussed, first an overview of available bondable strain gauges, namely metal film strain gauges and semiconductor strain gauges, is given. Metal film strain gauges consist of an alloy, predominantly copper-nickel (Constantan) or nickel-chromium (Karma), patterned on a flexible backing (e.g. polyimide) bonded by an epoxy glue to the surface of the structure under investigation. A photograph of a metal film strain gauge is shown in Fig. 3.13. The backing provides a means for handling the alloy pattern during installation and it also offers electrical insulation between the metal foil and the test structure. The patterned alloy forms a resistor of which the resistance value changes if the surface, to which the metal film strain gauge is bonded, experiences a deformation due to a load upon the test structure. The relation between the occurring strain (i.e. relative deformation) and the resistance change of a strain gauge is given by the gauge factor G (Eq. (3.13)), which varies between 2 and 5 for metal film strain gauges, dependent on the type of alloy used. Semiconductor strain gauges on the other hand depend on the piezoresistive property of silicon or germanium. Their resistivity changes as

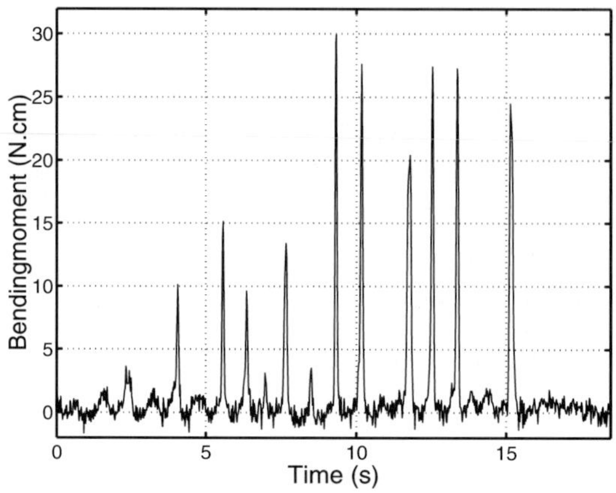

Figure 3.12: *Bending moment measurement, performed with the external measurement system, during chewing of bread.*

a result of stress occurring in the semiconductor. When the semiconductor experiences a stress σ (and therefore a strain $\sigma/E_{semiconductor}$), the lattice spacing between the atoms changes, affecting the band-gap energy. This change in band-gap energy either increases or decreases the number of available carriers, resulting in a resistance change. The gauge factor of a semiconductor strain gauge is dependent on the orientation of the resistor with respect to the semiconductor crystal lattice, the doping concentration and the type of dopant [Pue 93]. Two types of semiconductor strain gauges are commercially available: gauges consisting of a plain piece of semiconductor, usually bonded to a foil (e.g. phenolic glass), and gauges, consisting of a diffused region of impurities in a semiconductor crystal lattice.

Despite the fact that the gauge factor of semiconductor strain gauges is about two orders of magnitude larger than that of metal film strain gauges, the semiconductor strain gauge has been relegated to a small niche in today's total strain gauge market due to several reasons [Nag 01]. Because the metal is deposited onto polyimide or another flexible backing in the case of metal film strain gauges, they are easy to handle and use. Commercially available silicon strain gauges on the other hand are relatively small and extremely brittle. Metal film gauges have robust solder pads, while the lead wires of silicon gauges are often very small and connected to the gauge by conductive epoxy or ultrasonic means. Moreover, semiconductor strain gauges do not allow the same degree of flexibility in patterning as their metal film counterparts, which are sold in prearranged Wheatstone bridges, rosettes, and other patterns. Also the cost of semiconductor strain gauges is higher, compared with metal film strain gauges. This cost can be attributed to yield issues, repeatability, handling, lead attachment, testing requirements, and market size. Other important drawbacks of semiconductor strain gauges are their inferior linearity and higher temperature sensitivity.

Ideally, a strain gauge bonded to a test structure would respond only to the applied strain in

Figure 3.13: *Self temperature compensated metal film strain gauge with* R_0*=5kΩ and G=2.01 measuring 3.8 mm by 2.5 mm (FSM-A6306S-500-S6EC, BLH).*

the structure, and would be unaffected by other variables in the environment. Unfortunately, the electrical resistance of strain gauges varies not only with strain, but also with temperature. The temperature-induced resistance change of a strain gauge is given by [Pue 93]

$$\left(\frac{\Delta \mathrm{R}}{\mathrm{R}_0}\right)(\Delta \mathrm{T}) = (\alpha_R + G \cdot (\lambda_S - \lambda_R)) \cdot \Delta \mathrm{T} \tag{3.17}$$

with α_R the thermal coefficient of resistance of the resistive material of the strain gauge, λ_S the linear expansion coefficient of the structure to which the strain gauge is bonded, λ_R the linear expansion coefficient of the resistive material of the strain gauge, and ΔT the difference in temperature with the reference temperature T_0. The first contribution to the temperature-induced resistance change results from the temperature-dependent resistivity of the resistive material. The second term in Eq. (3.17) is the result of a difference in thermal expansion coefficients between the strain gauge's resistive material and the substrate material to which the gauge is bonded. The substrate expands or contracts due to a temperature change and since the strain gauge is firmly bonded to the substrate, the resistive material is forced to undergo the same expansion or contraction. If the thermal expansion coefficient of the strain gauge's resistive material differs from that of the substrate, this material is mechanically strained in conforming to the free expansion or contraction of the substrate. Since the resistive material is sensitive to this mechanical strain, the strain gauge resistance changes proportional to the difference in thermal expansion coefficients. It is clear from Eq. (3.17) that the thermal output depends not only on the nature of the gauge, but also on the material to which the gauge is bonded. Therefore, thermal output data are only meaningful when referred to a particular type of stain gauge, bonded to a specified substrate material.

One way to cope with temperature-induced resistance changes is the use of an *unstrained* dummy gauge, *identical* to the active strain gauge, subjected to the *same temperature* as the active gauge and mounted on the *same material*. By subtracting the temperature-induced resistance change of the dummy gauge from the resistance change of the active gauge, the stress-induced resistance change alone is obtained. Problems are encountered with this method of temperature compensation resulting from the difficulty to establish and maintain the above described *conditions*. Moreover, additional dummy strain gauges have to be used, requiring extra space, so that this temperature compensation method can not be applied for small objects as is the case for the *in vivo* measurements of the loads on oral implants. Another way to cope with temperature-induced resistance changes is the use of self temperature compensated strain gauges. If a strain gauge is bonded to a particular substrate material with a linear expansion coefficient λ_S, Eq. (3.17) shows that by selecting for the strain gauge a resistive material with appropriate thermal properties α_R and λ_R the temperature-induced resistance change can be limited. Self temperature compensated strain gauges have specially processed resistive materials with optimal thermal properties so that for a given substrate material the temperature-induced resistance change is restricted over a wide temperature interval. The resulting temperature-induced strain for a specific type of strain gauges and a specific substrate material, also called apparent strain ε_{app}, is given by

$$\varepsilon_{app} = A_0 + A_1\mathrm{T} + A_2\mathrm{T}^2 + A_3\mathrm{T}^3 + A_4\mathrm{T}^4 \qquad (3.18)$$

where T is the temperature and the coefficients A_i are obtained by least-squares approximation. This thermal output equation together with its tolerance are supplied by the manufacturer for each type of gauge. If the strain gauge is applied on a test substrate with a different thermal expansion coefficient $\lambda_{S_{Test}}$ from the one it is intended for, namely $\lambda_{S_{Ref}}$, the resulting apparent strain can be approximated with

$$\varepsilon_{app} = A_0 + A_1\mathrm{T} + A_2\mathrm{T}^2 + A_3\mathrm{T}^3 + A_4\mathrm{T}^4 + (\lambda_{S_{Test}} - \lambda_{S_{Ref}}) \cdot \Delta\mathrm{T} \qquad (3.19)$$

with ΔT the difference in temperature from the reference temperature T_0, i.e. the temperature of zero apparent strain. This means that the resulting thermal output curve is altered by a rotation of the original curve around the reference temperature. When the reference material thermal expansion coefficient is lower than the test material thermal expansion coefficient, the rotation is counterclockwise. When higher, the rotation is clockwise. Rotation of the thermal output curve by intentionally mismatching the test material thermal expansion coefficient and the reference thermal expansion coefficient can be employed to bias the thermal output characteristics so as to favor a particular working temperature range.

Table 3.2 and Table 3.3 give an overview of the properties of available strain gauges, which can be used for the *in vivo* measurements of the loads on oral implants. Table 3.2 gives an overview of the selected metal film strain gauges and Table 3.3 of suitable semiconductor strain gauges as far as their dimensions are concerned. Note that n-type silicon is employed to achieve temperature compensation for silicon strain gauges. Despite the many temperature compensation schemes developed for silicon over the years, self temperature compensated silicon strain gauges still have an inferior temperature behavior in comparison with self temperature compensated metal film strain gauges [Nag 01], which is an important drawback of silicon strain gauges. Silicon strain gauges also have an inferior linearity in comparison with metal film gauges. It is possible to correct for this non-linear behavior and for the temperature dependent behavior, but the latter requires the use of additional temperature sensors. Another drawback of silicon gauges

Code	FLG-02-11	FSM-A6306S-500-S6EC
Manufacturer	TML	BLH
Resistive material	Constantan	Stabiloy
Carrier	Epoxy	Polyimide-glass
Size	3.5 mm x 2.5 mm	3.8 mm x 2.5 mm
R_0	120 Ω	5000 Ω
Tolerance R_0	± 0.25 %	± 0.6 %
G	2.05	2.01
Tolerance G	± 1 %	± 0.5 %
Apparent strain	STC	STC
Tolerance STC	± 0.85 μstrain/°C	± 1.8 μstrain/°C
Temperature dependence G	0.01 %/°C	-0.011 %/°C
Safe bending radius	N/A	3 mm
Transverse sensitivity	3.7 %	-1.4 %
Strain range	3 %	2 %
Linearity	Linear	Linear

Table 3.2: Overview of the properties of the selected metal film strain gauges (STC = self temperature compensated).

Code	LN-100	SP5-09-100	SNB2-06-12S6	ESU-025-1500
Manufact.	BFGoodrich	BLH	BLH	Entran
Resist. mat.	p-type silicon	p-type silicon	n-type silicon	p-type silicon
Carrier	Polyimide	None	Phenolic glass	None
Size	2.5 x 2 mm^2	3.1 x 0.23 mm^2	3.1 x 2.54 mm^2	1.27 x 0.38 mm^2
R_0	5000 Ω	1000 Ω	120 Ω	1500 Ω
Tolerance R_0	N/A	± 3 %	± 10 %	± 20 %
G	85 to 140	145	-103	155
Tolerance G	N/A	± 10 %	± 10 %	± 5 %
App. strain	15.5 μstrain/°C	32.5 μstrain/°C	STC	35.1 μstrain/°C
Temp. dep. G	N/A	N/A	N/A	-0.32 %/°C
Safe bend. rad.	3 mm	76 mm	76 mm	N/A
Transv. sens.	-20 %	N/A	N/A	N/A
Strain range	3 %	< 0.5 %	< 0.5 %	< 0.3 %
Linearity	Linear	Non-linear	Non-linear	N/A

Table 3.3: Overview of the properties of available semiconductor strain gauges (STC = self temperature compensated). The apparent strain is referred to a carbon steel (1018) substrate ($\lambda_S = 12.06\ \mu$m/m/°C).

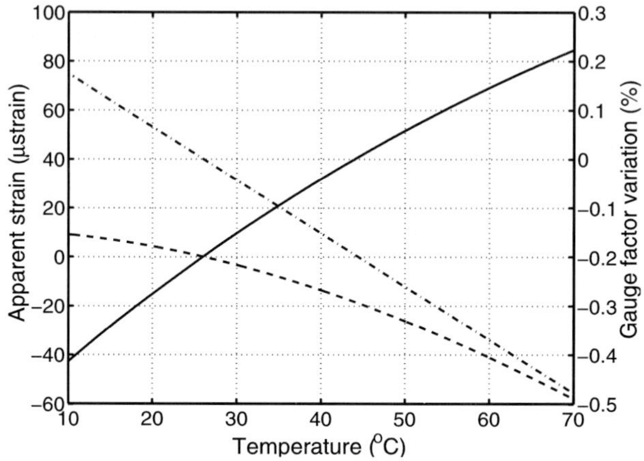

Figure 3.14: *Solid line: thermal output for a carbon steel (1018) substrate. Dashed line: thermal output for a titanium substrate. Dash-dot line: temperature dependence of G.*

is their inflexibility so that they can not be applied on rounded surfaces as is the case for oral implants. An exception to this is the LN-100 strain gauge, a relatively new strain gauge, fabricated based on micromachining techniques, but this type of strain gauge is not yet available in a self temperature compensated implementation. Because of the inferior temperature dependence, the inferior linearity, and the stiff character of silicon strain gauges, metal film strain gauges are chosen for the *in vivo* load measurements despite the much higher gauge factor G of silicon gauges. The FLG-02-11 type of strain gauge has been chosen for the external measurement system [Lie 99] and the FSM-A6306S-500-S6EC type for the miniaturized datalogger. A photograph of this strain gauge is shown in Fig. 3.13. This type of self temperature compensated strain gauge has been chosen, because of its high nominal resistance and yet relatively small dimensions, allowing a low power consumption for the battery-operated miniaturized datalogger. Moreover, system optimization proved that the introduced larger white noise due to the use of a high-resistance gauge is not the major noise contribution to the total system, as explained in Section 4.3.4.7. Although a limited safe bending radius of 3 mm is reported in the datasheets for the FSM strain gauge family, the manufacturer ensured that the abutments' radius of 2.25 mm would not impose a problem for the particular type of strain gauge, which has been confirmed during the measurements and the placement.

Fig. 3.14 shows the thermal properties of the FSM-A6306S-500-S6EC type of strain gauge, which is self temperature compensated for application on a carbon steel ($\lambda_S = 12.06\,\mu\text{m/m/}^{\text{o}}\text{C}$) substrate. The solid line in Fig. 3.14 gives the apparent strain equal to (T in ^{o}F)

$$\varepsilon_{app_{steel}} = [-136.3 + 2.14\text{T} - 5.79\text{E}^{-3}\text{T}^2 + 7.84\text{E}^{-6}\text{T}^3 - 5.95\text{E}^{-9}\text{T}^4]\,\mu\text{m/m} \qquad (3.20)$$

if the gauge is used on this type of substrate. When the gauge is used on a titanium ($\lambda_S = 8.82\,\mu\text{m/m/}^{\text{o}}\text{C}$) substrate, as is the case for oral implants, the resulting apparent strain can be calculated with Eq. (3.19) (T_0 = 78.8 °F (26 °C)) and is given by (T in °F)

$$\varepsilon_{app_{titanium}} = [5.54 + 0.34\mathrm{T} - 5.79\mathrm{E}^{-3}\mathrm{T}^2 + 7.84\mathrm{E}^{-6}\mathrm{T}^3 - 5.95\mathrm{E}^{-9}\mathrm{T}^4]\ \mu\mathrm{m/m} \qquad (3.21)$$

shown by the dashed line in Fig. 3.14. Also the gauge factor G itself varies with temperature, illustrated by the dash-dot line in Fig. 3.14. The slope of this line is equal to -0.011 %/°C. In order to be able to fully compensate the temperature-induced errors in case of the *in vivo* measurements of the loads on oral implants it is necessary to measure the local temperatures in the direct neighborhood of all the strain gauges and to correct the measured strains with calibrated thermal output data for each individual gauge. Since for a complete compensation this would require (at least) 18 additional temperature sensors with dedicated read-out and calibration circuits and since the expected temperature-induced errors, discussed further, are relatively small in comparison with the desired accuracy, this approach has been rejected. Moreover, one must keep in mind that the available space around the abutments is limited, because the original non-instrumented abutments are replaced by isolated instrumented abutments equipped with gauges, having a certain thickness. Problems can arise when instrumented abutments are installed into the original cavities in the gums due to their thickness. This is also a reason not to use temperature sensors, since these sensors and their connections would also contribute to the overall diameter of the instrumented abutments which may cause problems.

3.6 Specifi cations of the new miniaturized datalogger

The new miniaturized datalogger, part of the prosthesis, must be able to monitor up to 18 strain-gauge channels, corresponding with 6 abutments, during a two-day period. Because the monitoring has to go on continuously during this period, independent of the hospital environment, the datalogger is powered by two 1.55-V 41-mAh batteries [Ene]. This results in a low-power-design requirement for the autonomous datalogger. The maximum bandwidth of the *in vivo* load signals is 50 Hz, which has been experimentally verified with the external measurement system, requiring a minimum sample frequency of 100 Hz in each strain-gauge channel.

The prosthesis, equipped with the datalogger, and the strain gauges are protected by the patient's cheek, tongue and lip, causing the operating temperature of the datalogger and the strain gauges to be approximately equal to the patient's body temperature. In [Sun 02] a mean oral-temperature of 36.4 °C with a range between 33.2 °C and 38.2 °C (2σ-approach) has been measured in a group of 2749 individuals. The temperature range is different for men and women: for men a range between 35.7 °C and 37.7 °C is described and for women a range between 33.2 °C and 38.1 °C. The normal variation of the oral temperature equals 0.5-0.7 °C, i.e. the normal daily body-temperature fluctuation [Gan 95]. The effect of hot beverages, cold beverages, and chewing gum on the oral temperature also has been investigated. In [New 01] the maximum mean-oral-temperature shift caused by these actions is reported to be lower than ±1.5 °C. Because the datalogger is embedded within the prosthesis, protected by the patient's cheek, tongue and lip, it may be expected that the variation of the datalogger's operating temperature due to these actions is very small, especially since they normally only last for a short period of time.

The measurement accuracy $\varepsilon_{\mathrm{err},\sigma}$ in each channel, defined as the maximum standard deviation of the measurement error, is 10 μstrain. This corresponds with a maximum standard deviation of the error $\sigma_{\mathrm{R}(S_{\mathrm{i}})}$ for the strain gauge resistance ($\mathrm{R}(S_{\mathrm{i}})$) measurements of 100.5 mΩ (Eq. (3.13)). The strain gauge resistance measurement values are independent of each other so that the axial-force and bending-moment measurement accuracies ($\sigma_{\mathrm{F_{ax}}}$, $\sigma_{\mathrm{M_x}}$ and $\sigma_{\mathrm{M_y}}$) can be derived from Eq. (3.2), Eq. (3.15) and Eq. (3.4) with the following formula [Abr 72, Bev 69]:

$$\sigma_z^2 \sim \sigma_{\mathrm{R}(S_1)}^2 \left(\frac{\delta z}{\delta \mathrm{R}(S_1)}\right)^2 + \sigma_{\mathrm{R}(S_2)}^2 \left(\frac{\delta z}{\delta \mathrm{R}(S_2)}\right)^2 + \sigma_{\mathrm{R}(S_3)}^2 \left(\frac{\delta z}{\delta \mathrm{R}(S_3)}\right)^2 \tag{3.22}$$

where

$$z = \mathrm{F_{ax}}, \mathrm{M_x} \text{ or } \mathrm{M_y} \tag{3.23}$$

These equations yield

$$\sigma_{\mathrm{F_{ax}}} \sim \sqrt{\frac{1}{3}}\,\sigma_{\mathrm{R}(S_{\mathrm{i}})} = 58\ \mathrm{m\Omega} \tag{3.24}$$

and

$$\sigma_{\mathrm{M_x}} = \sigma_{\mathrm{M_y}} = \sigma_{\mathrm{M_{x,y}}} \sim \sqrt{\frac{2}{3}}\,\sigma_{\mathrm{R}(S_{\mathrm{i}})} = 82\ \mathrm{m\Omega} \tag{3.25}$$

corresponding with a measurement accuracy for the axial force $\mathrm{F_{ax}}$ of 5.6 N and a measurement accuracy for the bending moment load components $\mathrm{M_{x,y}}$ of 0.64 N.cm (cf Table 3.1). By combining Eq. (3.15), Eq. (3.4), Table 3.1, Eq. (3.5) and Eq. (3.22), the accuracy σ_{M} of the overall bending-moment M measurement can be derived:

$$\sigma_{\mathrm{M}} = \sqrt{\frac{2}{3}} \cdot \frac{\sigma_{\mathrm{R}(S_{\mathrm{i}})}}{127.21\ \frac{\mathrm{m\Omega}}{\mathrm{N.cm}}} = 0.64\ \mathrm{N.cm} \tag{3.26}$$

In vivo load measurements, carried out with the external measurement system, have shown that the maximum load that can occur due to excessive biting is equal to a combination of an axial force $\mathrm{F_{ax}}$ of ±600 N (compression and tension) and a bending moment M of ±105 N.cm (positive and negative). This combined maximum load gives rise to a maximum/minimum strain $\varepsilon_{\mathrm{max,min}}$ of ±1948 μstrain (compressive and tensile) which corresponds with a strain gauge resistance change of ±19.6 Ω. The tolerance on the strain gauges' nominal resistance equals ±0.6 % (Table 3.1) which is equivalent to a nominal-resistance variation of 30 Ω. Because this is 1.5 times as big as the resistance change due to the maximum/minimum strain, compensation of every strain-gauge channel is required at the beginning of the measurements. Moreover, when the abutments and the prosthesis are placed orally, it has been found with the external measurement system that an excessive pre-strain can occur due to mechanical misalignments between the individual implant-abutment entities and the prosthesis bridging structure. This pre-strain can be as big as ±3232 μstrain caused by a combined axial force $\mathrm{F_{ax}}$ of ±1600 N and a bending moment M of ±125 N.cm. This pre-strain is thus also bigger than the maximum/minimum occurring strain due to excessive biting, which shows that the new measurement system must be capable of compensating for this excessive pre-strain and the strain-gauge nominal-resistance tolerance *after* placement of the prosthesis instrumented with the datalogger.

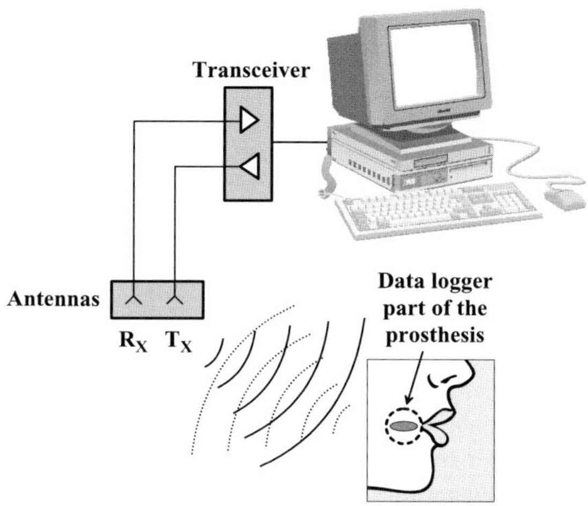

Figure 3.15: *Overview of the complete bi-directional telemetry system.*

For the external measurement system this compensation is carried out manually by means of potentiometers (cf Fig. 3.10) for each individual channel. The new miniaturized datalogger on the other hand is able to compensate *itself* for the offsets introduced in the strain-gauge channels. The compensation is carried out automatically by commanding the datalogger wirelessly to compensate towards a user-definable output value for a selectable strain-gauge channel. To achieve this a digital automatic-compensation block and a bi-directional transceiver, both described in detail in Chapter 5, are integrated in the datalogger. The integrated bi-directional transceiver permits wireless activation of the compensation *after* placement of the prosthesis. Moreover, it allows to (re-)program the datalogger's operation-mode settings (e.g. the number of strain-gauge channels, the data processing algorithm, ...). By the incorporation of flexibility into the operation of the datalogger and the possibility of reprogramming the device settings wirelessly, a highly-flexible autonomous datalogger is obtained, which can be tailored towards each individual patient *in situ*.

Fig. 3.15 shows an overview of the complete bi-directional telemetry system. The datalogger's internal transceiver communicates with an external RF unit connected to a PC. Dedicated software runs on the PC to program/read out the datalogger and to store and visualize the collected data. The datalogger is a transponder-type device: it is able to pick up a nearby low-frequency programming/reading field (132 kHz) and respond to it. The datalogger can be (re-)programmed by this field or can be instructed to send the collected data to the external RF unit. The datalogger can also be commanded to send its status bytes so that the actual operation mode of the device can be retrieved. The receiver R_x and transmitter T_x antennas of the external RF unit are LC-circuits tuned respectively to 66 kHz and 132 kHz. By amplitude modulation of the 132-kHz carrier, transmitted by the RF unit, the datalogger can be (re-)programmed. On

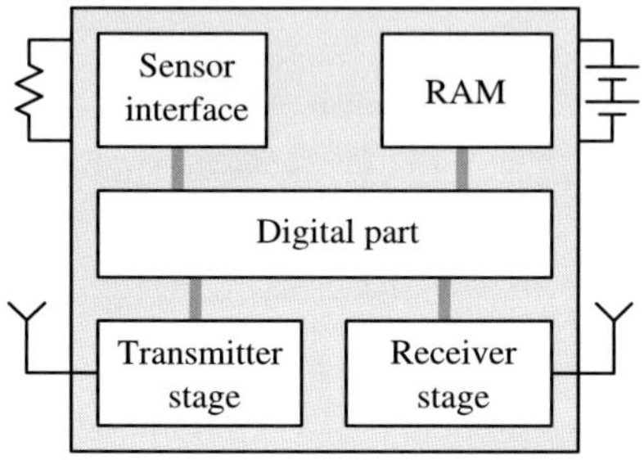

Figure 3.16: *Overview of the complete datalogger.*

the other hand data are transmitted from the datalogger to the receiver antenna R_x by modulation of a 66 kHz-carrier, which is derived on board from an incoming non-modulated 132-kHz field, generated by the transmitter antenna T_x.

Fig. 3.16 shows an overview of the complete datalogger . It consists of 4 major parts: a multi-gauge sensor interface with digitally-programmable offset-compensation, a digital part adding intelligence to the datalogger, a wireless transceiver for communication with the external world and a memory for storage of the measured data. The datalogger is powered by two 1.55-V batteries in series providing a supply voltage of 3.1 V. The development of the complete datalogger has been done in two stages. First the sensor interface has been integrated into a single chip (cf Chapter 4). In a second stage a single-chip datalogger IC, including the sensor interface, the digital part, and the wireless transceiver has been developed (cf Chapter 5). Both chips are fabricated in a standard 0.7-μm CMOS technology (Alcatel Microelectronics C07MA).

3.7 Conclusion

In this chapter the system concept of the miniaturized datalogger for stress monitoring in oral implants has been presented. Titanium implants are used to support/retain dental prostheses, based on the principle of osseointegration. Despite their optimistic success rate, failures do occur and originate from mechanical and/or biological causes. The latter are generally characterized by the loss of osseointegration or continuous marginal bone loss. Several clinical studies have put forward unfavorable loading conditions as the most important cause of biological implant failures. Nonetheless, no quantitative *in vivo* implant/bone load data are available to investigate the processes involved in bone remodeling and implant failures.

To measure these loads, strain gauge sensors are installed on the abutments. These are (titanium) cylinders, located in the gums, which are placed on top of the implants. Every abutment is equipped with 3 strain gauges, placed 120° from each other. By combining the measured strain gauge resistance values, the axial force F_{ax} and the bending moment M, imposed on the abutment, can be derived.

A first important step in the measurement of *in vivo* loads has been the development of an

<table>
<tr><td>Monitoring period</td><td colspan="2">2 days</td></tr>
<tr><td>Power supply (batteries)</td><td colspan="2">3.1-V 41-mAh</td></tr>
<tr><td>Number of channels</td><td colspan="2">18</td></tr>
<tr><td>Bandwidth</td><td colspan="2">50 Hz</td></tr>
<tr><td>Mean oral-temperature</td><td colspan="2">36.4 °C</td></tr>
<tr><td>Oral-temperature variation</td><td colspan="2">0.5-0.7 °C</td></tr>
<tr><td rowspan="4">Measuring accuracy (σ)</td><td>$\varepsilon_{\mathrm{err},\sigma}$</td><td>10 μstrain</td></tr>
<tr><td>$\mathrm{R}(S_\mathrm{i})\sigma$</td><td>100.5 m$\Omega$</td></tr>
<tr><td>F_{ax}</td><td>5.6 N</td></tr>
<tr><td>M</td><td>0.64 N.cm</td></tr>
<tr><td rowspan="2">Maximum/minimum strain
(excessive biting)</td><td>$\varepsilon_{\mathrm{max,min}}$</td><td>$\pm$ 1948 μstrain</td></tr>
<tr><td>$\Delta\mathrm{R}(S_\mathrm{i})_{\mathrm{max,min}}$</td><td>$\pm$ 19.6 Ω</td></tr>
</table>

Table 3.4: Summary of the datalogger specifications.

external non-portable measurement system. However, this external system has several drawbacks. The strain gauges are connected to the external measurement unit by wires, which have to come out of the patient's mouth and may introduce artificial chewing behavior. Also the fact that the measurements are done in the hospital and on command may introduce measurement artifacts. Moreover, unconscious day as well as nocturnal dental activities, like bruxing and clenching, seen as a missing link for the validation of existing bone remodeling models, can not be measured. These drawbacks clearly demonstrate the need for the presented miniaturized (battery-operated) datalogger, part of the prosthesis, capable of measuring continuously over a longer period independent of the hospital environment.

A comparison between semiconductor and metal film strain gauge sensors has been made. Despite their larger gauge factor semiconductor strain gauges are relegated to a small niche in comparison to metal film strain gauges due to their inflexibility, higher cost, inferior linearity and higher temperature sensitivity. To limit temperature-induced errors, self temperature compensated metal film strain gauges are applied in this work. The temperature-induced apparent strain for these strain gauges is restricted if they are applied on a specific substrate material; titanium in this case.

Because of the offsets introduced in the different strain-gauge channels by the tolerance on the strain gauges' nominal resistance and potential pre-strains resulting from misalignments between the prosthesis and the implants/abutments, compensation for every individual strain-gauge channel *after* placement of the prosthesis is required. Therefore, a bi-directional transceiver is included in the datalogger, which is used to activate the implemented automatic offset-compensation for a given strain-gauge channel. Moreover, it allows to (re-)configure the datalogger settings, like e.g. its data processing algorithm. The new datalogger consists of 4 major parts: a multi-gauge sensor interface, a digital part adding intelligence, a wireless transceiver and a memory. To conclude the specifications for the datalogger are summarized in Table 3.4.

Chapter 4

Multi-gauge offset-compensated sensor interface chip

4.1 Introduction

To deal with the complexity of the total datalogger system, described in the previous chapter, its implementation has been carried out in two steps. First, the sensor interface has been integrated on a separate sensor interface chip. In a second stage, a single-chip datalogger IC, implementing the complete datalogger system with exception of the memory, has been realized. This chapter presents the design of the sensor interface chip. It is able to measure up to 18 strain-gauge channels and contains digitally-programmable offset-compensation for every channel to cope with potential offsets due to pre-strains and the tolerance on the strain gauges' nominal resistance. The complete datalogger chip, including the sensor interface, the digital part and the wireless transceiver, is described in the following chapter. Special care has been taken to restrict the power consumption of the chips.

First, the measurement setup of the sensor interface chip is discussed in this chapter. A current-driven Wheatstone configuration is applied to interface with the strain gauges. To cope with potential offsets, a compensation setup, consisting of a current-steering DAC, a digital interface and an on-chip memory to store the required digital compensation words, is implemented. The total sensor interface chip includes a reference current source, an 8-bit DAC, a digital interface, a compensation-words memory, a SC instrumentation amplifier, a SC S/H, a 9-bit successive approximation ADC and a relaxation clock oscillator. The design of these building blocks is described in detail. After illustrating the layout of the realized sensor interface chip the chapter ends with measurement results.

4.2 Measurement/compensation setup

To measure the resistance values of the strain gauges S_i a Wheatstone-bridge is implemented in the sensor interface chip. There are two different Wheatstone-bridge configurations: the voltage-driven one shown on the left in Fig. 4.1 and the current-driven one shown on the right. The former

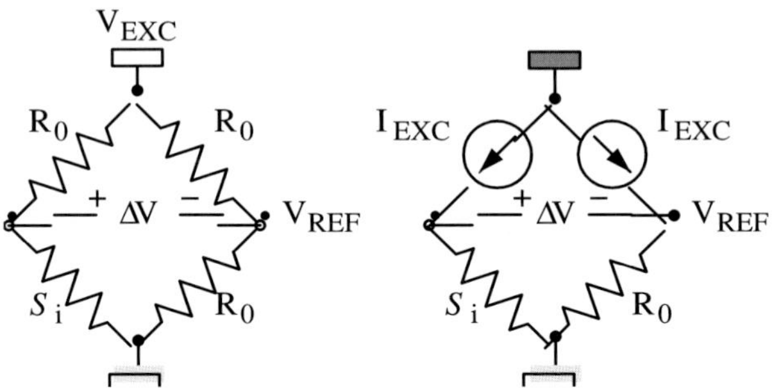

Figure 4.1: *Comparison between voltage-driven (left) and current-driven (right) Wheatstone configuration.*

consists of the strain gauge under measurement S_i, 3 resistors equal to the nominal resistance R_0 of S_i, and a voltage source V_{EXC}, while the latter consists of S_i, a reference resistor equal to R_0, and two current sources I_{EXC}. For the voltage-driven configuration the relation between the output voltage ΔV of the bridge and a strain gauge resistance change $\Delta R(S_i)$ due to a strain ε is given by Eq. (3.16):

$$\Delta V = \frac{\Delta R(S_i) \cdot V_{EXC}}{2 \cdot (2 \cdot R_0 + \Delta R(S_i))} \approx \frac{1}{4} \cdot G \cdot \varepsilon \cdot V_{EXC} \tag{4.1}$$

while for the current-driven configuration this is given by

$$\Delta V = \Delta R(S_i) \cdot I_{EXC} = G \cdot \varepsilon \cdot R_0 \cdot I_{EXC} \tag{4.2}$$

The ratio between the output voltage ΔV and the total current consumption I_{total} for a strain ε is equal to Eq. (4.3) and Eq. (4.4) respectively for the voltage-driven and the current-driven configuration.

$$\frac{\Delta V}{I_{total}} = \frac{\Delta V}{\frac{V_{EXC}}{R_0}} = \frac{1}{4} \cdot G \cdot \varepsilon \cdot R_0 \tag{4.3}$$

$$\frac{\Delta V}{I_{total}} = \frac{\Delta V}{2\, I_{EXC}} = \frac{1}{2} \cdot G \cdot \varepsilon \cdot R_0 \tag{4.4}$$

These equations show that the current-driven configuration (Eq. (4.4)) has a doubled sensitivity compared to the voltage-driven one (Eq. (4.3)) so that a lower power consumption can be achieved with it. This is the reason why the current-driven configuration has been selected for implementation. Another advantage of this configuration is its linear response to $\Delta R(S_i)$ which is not the case for the voltage-driven one, because of the $\Delta R(S_i)$-term in the denominator of Eq. (4.1).

Fig. 4.2 shows how the current-driven configuration is extended with a digitally-controllable compensation current I_{DAC} to cope with the offset introduced by the nominal-resistance tolerance of S_i and the pre-strain. I_{SOURCE} (not shown) is a reference current source from which the

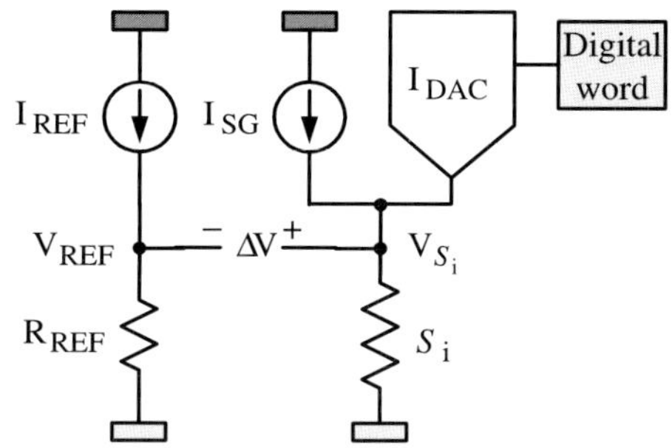

Figure 4.2: *Compensation setup for the current-driven Wheatstone configuration.* $R_0(S_i)$*=5000 Ω and* R_{REF}*=9310 Ω. The current values are listed in Table 4.1.*

currents I_{REF}, I_{SG} and the digitally-controllable current I_{DAC} are derived. The current I_{REF} flows through the reference resistor R_{REF} yielding a reference voltage V_{REF}. To reduce the power consumption R_{REF} is chosen larger than $R_0(S_i)$. Note that the upper limit of R_{REF} is determined by settling and noise considerations (cf Section 4.3.4.4 and Section 4.3.4.7). The total current I_{SG}+I_{DAC} through the strain gauge S_i is digitally-controllable. By applying the appropriate digital word at the input of I_{DAC}, which is implemented as a binary-weighted current-steering DAC, the current through S_i can be adjusted so that the voltage V_{S_i} across S_i becomes equal (within the resolution level of the DAC) to V_{REF} and compensation is achieved. After compensation the output voltage $\Delta V = V_{S_i} - V_{REF}$ for a strain ε is given by

$$\Delta V = [I_{SG} + I_{DAC}]\, R_0(S_i) + [I_{SG} + I_{DAC}]\, G\,\varepsilon\, R_0(S_i) - I_{REF}\, R_{REF} \quad (4.5)$$

$$\approx [I_{SG} + I_{DAC}]\, G\,\varepsilon\, R_0(S_i) \approx G\varepsilon V_{REF} \quad (4.6)$$

which is (within the resolution level of the DAC) independent of the resistance value $R_0(S_i)$ of S_i and the pre-strain.

Fig. 4.3 illustrates the system expanded for 18 strain gauges. It shows the introduction of multiplexers (MUX) to switch between the 18 different strain-gauge channels. Compensation for every strain-gauge channel separately is carried out after placement of the prosthesis. The digital words needed for compensation are programmed into an on-chip nulling memory REG, composed of registers, using the PROG/SEL-block. In the measurement mode, when a particular strain-gauge channel is measured, the digital word belonging to that particular channel is fetched from REG and offered to the DAC so that offset-compensated measurements are performed.

The output voltage ΔV of the multi-gauge measurement setup is amplified by a switched-capacitor amplifier described in Section 4.3.4. This amplifier samples the output voltages ΔV of the 18 different strain-gauge channels consecutively at 2 kHz so that the sampling frequency of each channel equals 111 Hz satisfying the minimum sample-frequency requirement of 100 Hz. The selection of a strain-gauge channel is done by applying its 5-bit channel-number at the input of the PROG/SEL-block as discussed further. To reduce the power consumption a special clock ϕ_{sample} (Fig. 4.3) is employed to clock the current sources I_{REF}, I_{SG} and I_{DAC} switching off these currents during most of the time of the switched-capacitor amplifier's reset phase.

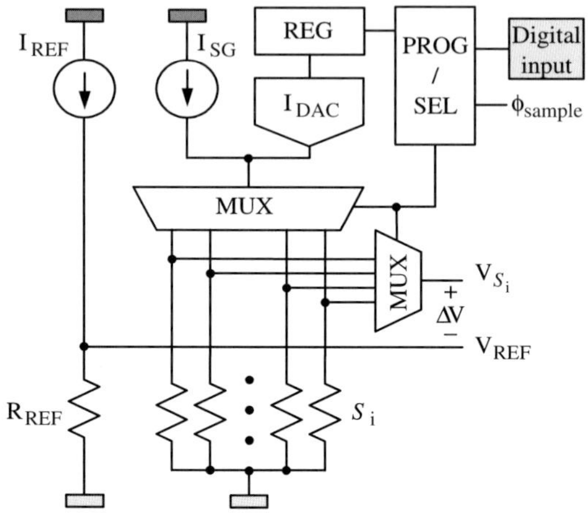

Figure 4.3: *Compensation setup for 18 strain gauges.*

4.3 Sensor interface building blocks

4.3.1 Reference current source

4.3.1.1 Operating principle

Fig. 4.4 shows the reference current source I_{SOURCE} from which I_{REF}, I_{SG} and I_{DAC} (Fig. 4.3) are derived. Q1 and Q2 consist of vertical pnp bipolar transistors [San 94]. The realization of these transistors in the Alcatel Microelectronics standard 0.7-μm *n*-well CMOS technology is shown in Fig. 4.4 (not drawn to scale). A p^+-region inside the *n*-well, corresponding with a PMOS source/drain region, serves as the emitter of the bipolar transistor, the *n*-well as the base and the *p*-type substrate connected to V_{SS} (= 0 V) as the collector. Q1 consists of a single unit bipolar transistor and Q2 of K unit bipolar transistors in parallel. The model (VPNP460) and the dimensions of the employed unit vertical pnp bipolar transistor are given in [Alc 97, Alc 01]. The relation between the base-emitter voltage V_{BE} and the collector current I_C is given by [San 94]

$$V_{BE} = V_T \ln\left(\frac{I_C}{I_S}\right) \tag{4.7}$$

with $V_T = \frac{kT}{q}$ the thermal voltage and I_S the saturation current.

The transistors M1-M4 form a low-output-voltage (wide-swing) cascode current mirror. The maximum allowable voltage at the drain of M3 equals V_{DD} (= 3.1 V) $- |V_{DSsat1}| - |V_{DSsat3}|$. To achieve this the gate voltage of M3/M4 is biased at $V_{DD} - |V_{DSsat1}| - |V_{DSsat3}| - |V_{T3}|$. For the implemented circuit a safety margin of 0.15 V is included, resulting in a gate voltage of M3/M4 of 1.6 V. The biasing of this voltage is done with the transistors M7-M10 [Joh 97].

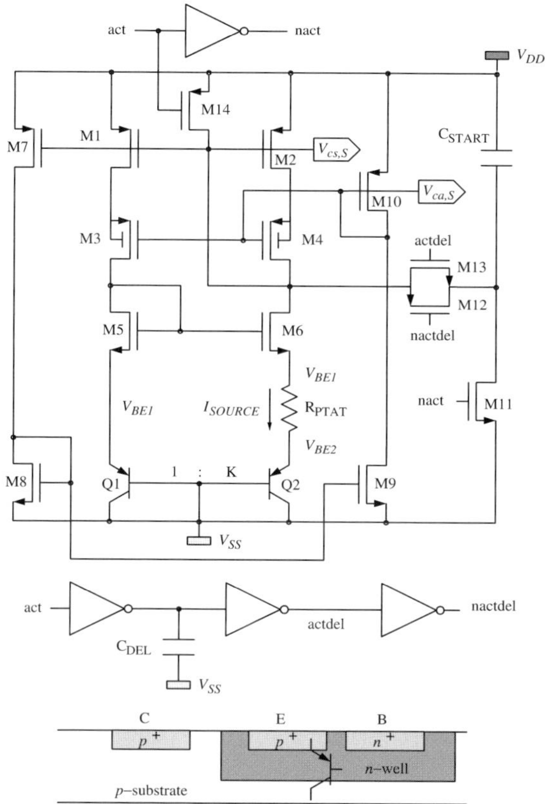

Figure 4.4: *Self-biased thermal-voltage-referenced reference current source (K=8). Component dimensions are listed in Table A.1.*

The current mirror M1-M4 ensures that the currents through Q1 and Q2 are equal. Because of the same current through M5 and M6 and because their gates are connected to each other, their source voltages are identical (neglecting channel-length modulation). As a result the current I_{SOURCE} through the resistor R_{PTAT} equals approximately (cf Eq. (4.7))

$$I_{SOURCE} = \frac{\Delta V_{BE}}{R_{PTAT}} = \frac{V_{BE1} - V_{BE2}}{R_{PTAT}} = \frac{V_T \ln\left(\frac{I_{SOURCE}\cdot\beta}{I_S\cdot(\beta+1)}\right) - V_T \ln\left(\frac{I_{SOURCE}\cdot\beta}{K\cdot I_S\cdot(\beta+1)}\right)}{R_{PTAT}} = \frac{V_T \ln(K)}{R_{PTAT}} \quad (4.8)$$

with β ($\approx$ 22.5) the current gain of the vertical bipolar transistors. This equation reveals that I_{SOURCE} is proportional to absolute temperature (PTAT). Eq. (4.8) also illustrates that the reference current source is self-biased: the reference voltages V_{BEi} determining the current I_{SOURCE} depend on I_{SOURCE} itself. This property gives rise to a degenerated bias point [Raz 01]. Besides the wanted stable operating point, the circuit is also stable if all the transistors carry a zero current. To solve this problem a start-up circuit that drives the circuit out of its degenerated bias point is added. It is activated by a low-to-high transition of *act* (Fig. 4.4). When *act* is low, C_{START} is charged to the supply voltage V_{DD}-V_{SS}. During activation, the terminal of C_{START} connected to V_{SS} is switched to the gates of M1-M2 after switching off first M11 and M14. Due to gate-source voltage imposed in this way the transistors M1 and M2 start to inject current, discharging C_{START}, and the circuit evolves towards the desired operating point.

MAPLE has been used to optimize the overall system performance with emphasis on the power consumption taking into account the different design criteria/equations discussed in the remainder of this chapter. This optimization yields a current reference source I_{SOURCE} equal to 22 μA. The other currents found by optimization are summarized in Table 4.1. The digitally-controllable current I_{DAC} is described in detail in Section 4.3.2. The resistance of the reference resistor R_{REF} (Fig. 4.3) equals 9310 Ω.

4.3.1.2 Accuracy and mismatch

The accuracy of I_{SOURCE} is determined by the accuracy of R_{PTAT}, the offset between the sources of M5 and M6, and the mismatch between the unit elements of Q1 and Q2. The resistor R_{PTAT} is an (externally) screen-printed resistor. Its resistance value can be trimmed to the wanted value ($\approx$ 2777 Ω) with an accuracy of 0.1 %. This resistor accuracy and the related current reference accuracy [Opt 86] can not be achieved with a (non-trimmable) on-chip resistor. The best absolute accuracy of a resistor in the Alcatel Microelectronics C07MA technology is $\pm$7.7 %.

In order to find the critical design parameters for the random offset between the sources of M5 and M6 the relative error of the current, respectively through M1/M2 and M5/M6, is calculated [Pie 98]. The relative error of the current through M1/M2 (subscripts p) due to mismatch is given by [Bas 98]

$$\frac{\Delta I}{I} = \frac{\Delta \beta_p}{\beta_p} + \frac{2\Delta V_{T,p}}{V_{GS,p} - V_{T,p}} \tag{4.9}$$

On the other hand the relative error of the current through M5/M6 (subscripts n) due to mismatch is given by

$$\frac{\Delta I}{I} = \frac{\Delta \beta_n}{\beta_n} + \frac{2\Delta V_{T,n}}{V_{GS,n} - V_{T,n}} + \frac{2\Delta V_{GS,n}}{V_{GS,n} - V_{T,n}} \tag{4.10}$$

Current	Value	M
I_{SOURCE}	22 μA	1
I_{SG}	308 μA	14
I_{DAC}	(0..255)x172 nA	(0..255)x1/128
I_{REF}	176 μA	8

Table 4.1: Overview of the currents with $M = I_i/I_{SOURCE}$.

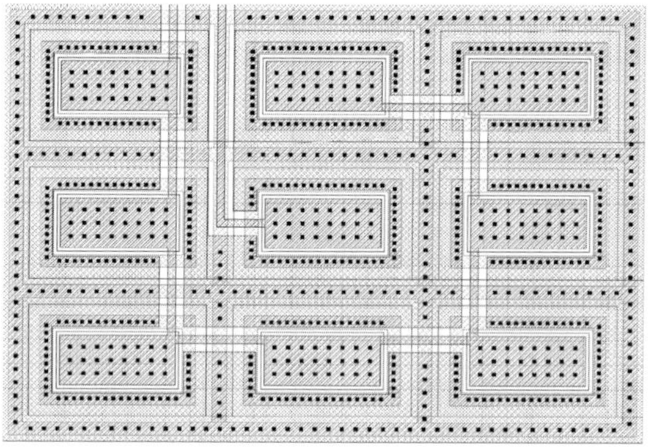

Figure 4.5: *Centroid layout of Q1 (in the middle) and Q2 (K=8).*

Note the introduction of an extra extra term due to the difference in gate-source voltages ΔV_{GS} between M5 and M6. Since the relative error $\frac{\Delta I}{I}$ in Eq. (4.9) and Eq. (4.10) must be the same, the variance $\sigma^2(\Delta V_{GS})$ of the offset between the sources of M5 and M6 can be found (cf Eq. (3.22)):

$$\sigma^2(\Delta V_{GS}) = \frac{(V_{GS,n} - V_{T,n})^2}{4}\left(\sigma^2\left(\frac{\Delta\beta_p}{\beta_p}\right) + \sigma^2\left(\frac{\Delta\beta_n}{\beta_n}\right)\right) + \frac{(V_{GS,n} - V_{T,n})^2}{(V_{GS,p} - V_{T,p})^2}\sigma^2(\Delta V_{T,p}) + \sigma^2(\Delta V_{T,n}) \tag{4.11}$$

From Eq. (4.11) follows that $V_{GST,p}$ must be chosen large and $V_{GST,n}$ small. In the implemented circuit $V_{GST,p}$ and $V_{GST,n}$ are equal to 0.4 V and 0.2 V respectively. $\sigma^2(\Delta V_T)$ and $\sigma^2\left(\frac{\Delta\beta}{\beta}\right)$ are given by [Bas 98]

$$\sigma^2(\Delta V_T) = \frac{(A_{1VTO} + A_{1\gamma}(\sqrt{2\phi_F + V_{SB}} - \sqrt{2\phi_F}))^2}{W{\cdot}L} \approx \frac{A_{1VTO}^2}{W{\cdot}L} \tag{4.12}$$

$$\sigma^2\left(\frac{\Delta\beta}{\beta}\right) = \frac{A_\beta^2}{W{\cdot}L} \tag{4.13}$$

To calculate the random-offset value the C07MA-technology mismatch parameters, summarized in [Bas 98], have been used, resulting in a $\sigma(\Delta V_{GS})$ equal to 392 μV.

To reduce the mismatch between Q1 and Q2 they consist of unit bipolar transistors implemented in a centroid layout, depicted in Fig. 4.5 [San 94]. Q2 consists of 8 unit bipolar transistors placed symmetrically around Q1. ΔV_{BE} is approximately equal to 61 mV, which means that the error due to the random offset $\sigma(\Delta V_{GS})$ is limited to ± 0.64 %.

4.3.1.3 Supply-voltage dependence

The employed batteries are silver oxide batteries [Ene], which provide a stable operating voltage until the end of discharge, so that supply-voltage variations due to the batteries are limited. On the other hand, supply-voltage variations may also be introduced by the switching of the datalogger's digital part. To reduce the dependence of the reference current on the supply voltage the wide-swing cascode current mirror M1-M4 is implemented instead of a simple current mirror. M3 and M4 ensure that the drain-source voltages of M1 and M2 are approximately the same eliminating inaccuracies due to channel-length modulation [San 94].

The relation between a DC-variation of V_{DD}/V_{SS}, and the variation of I_{SOURCE} is found by small-signal analysis:

$$\frac{\delta I_{SOURCE}}{\delta V_{DD,SS}} \approx \frac{g_{o6}}{(g_{m6}+g_{mb6})\cdot(R-R_{Q1})} \approx -128.5 \text{ dB} \quad (4.14)$$

with $R=R_{PTAT}+R_{Q2}$. Eq. (4.14) reveals that the sensitivity can be reduced by making g_{m6} and r_{ds6} large or in other words, by making $V_{GST,n}$ small and L,n large. The simulated dependence of I_{SOURCE} for an (arbitrarily-chosen) supply-voltage interval of 0.4 V at a temperature of 36.4 °C is depicted in Fig. 4.6 (a). The relative error of I_{SOURCE} over the complete interval is smaller than $\pm$ 0.4 %. The solid line in Fig. 4.6 (b) gives the supply-voltage dependence of I_{SOURCE} as a function of the frequency. It can be seen that the sensitivity is lower than -119 dB up to a frequency of 1 MHz. The DC-sensitivity equals -127 dB, which corresponds with Fig. 4.6 (a).

The used silver oxide batteries [Ene] have a stable operating voltage until the end of discharge. The maximum variation from 1.55 V of the battery operating voltage is limited to $\pm$20 mV. This means that the total variation of the supply voltage (i.e. 3.1V) is limited to $\pm$40 mV, resulting in a maximum relative error $\frac{\Delta I_{SOURCE}}{I_{SOURCE}}$ over the supply-voltage interval equal to $\pm$0.08 %. From Eq. (4.5) follows that the resulting voltage difference ΔV as a function of a strain ε and the supply voltage difference ΔV_{dd} of $\pm$40 mV is given by

$$\begin{aligned}\Delta V(\Delta V_{dd}) &= (I_{SG}+I_{DAC})\cdot G\cdot\varepsilon\cdot R_0(S_i)\cdot(1+\frac{\Delta I_{SOURCE}}{I_{SOURCE}}) \\ &+ \left[(I_{SG}+I_{DAC})\cdot R_0(S_i) - I_{REF}\cdot R_{REF}\right]\cdot(1+\frac{\Delta I_{SOURCE}}{I_{SOURCE}})\end{aligned} \quad (4.15)$$

The first term in Eq. (4.15) is proportional to the strain ε to be measured. The maximum error occurs for the maximum/minimum strain $\varepsilon_{max,min}$ and equals approximately $\pm$1.58 μstrain. The second term in Eq. (4.15) is the residual offset (cf Eq. (4.29)) after compensation, determined by the resolution level and the accuracy of I_{DAC} (cf Section 4.3.2.1). The maximum error due to this term equals approximately $\pm$0.35 μstrain, so that the worst-case error due to the supply-voltage dependence of I_{SOURCE} equals approximately $\pm$1.9 μstrain.

Note the importance of the placement of the start-up capacitance C_{start}. Several start-up schemes are applicable for this circuit. The implemented one has the advantage that after start-up C_{start} forms a bypass capacitor between the gate and source (i.e. V_{DD}) of M1/M2 ensuring a lower supply-voltage dependence of the current mirror M1-M2 at 'higher' frequencies. The dashed line in Fig. 4.6 (b) shows the sensitivity for the case where C_{start} would have been placed between the gate of M1/M2 and V_{SS}. In this case the sensitivity at a frequency of 1 MHz is increased to -83 dB. At the digital part's operating frequency of 128 kHz, the sensitivity for this start-up scheme is -87 dB, which is about 36 dB worse compared with the sensitivity of implemented one (-123 dB). It is clear that this alternative start-up scheme must be rejected.

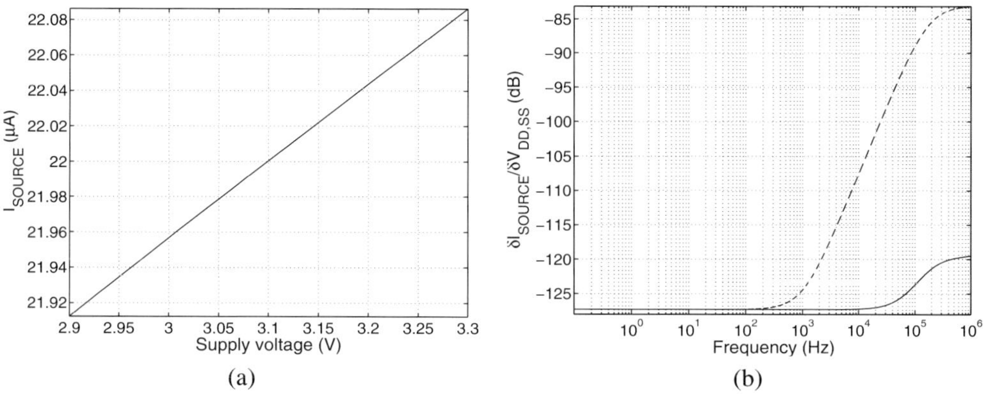

Figure 4.6: *(a) Supply-voltage dependence of I_{SOURCE} at 36.4 °C. (b) Frequency dependence of the supply-voltage sensitivity of I_{SOURCE}.*

4.3.1.4 Temperature dependence

The temperature dependence of I_{SOURCE} follows from Eq. (4.8):

$$\frac{1}{I_{SOURCE}}\frac{\delta I_{SOURCE}}{\delta T} = \frac{1}{V_T}\frac{\delta V_T}{\delta T} - \frac{1}{R_{PTAT}}\frac{\delta R_{PTAT}}{\delta T} = \frac{1}{T} - \frac{1}{R_{PTAT}}\frac{\delta R_{PTAT}}{\delta T} \tag{4.16}$$

$$= 3230.5 \text{ ppm/}^{o}\text{C} - \frac{1}{R_{PTAT}}\frac{\delta R_{PTAT}}{\delta T} \text{ at } 36.4\ ^{o}\text{C} \tag{4.17}$$

As explained in Section 5.9 the total system is embedded within the prosthesis, more in particular within the isolating material of the prosthesis. Therefore a homogeneous temperature distribution is expected for the chip and the Al_2O_3 hybrid carrying the datalogger chip and the other components. The hybrid also contains the screen-printed resistor R_{PTAT}. By selecting a resistor paste with a positive temperature coefficient of resistance (TCR) the two terms in Eq. (4.16) partially cancel each other. A resistor paste suitable for this application is the 5093D resistor-paste series from Dupont [Dup a]. It has a nominal sheet resistance of 1000 Ω/□ and a nominal TCR of 2750 ppm/°C. From Eq. (4.17) follows that the use of this resistor paste results in a temperature dependence of I_{SOURCE} equal to 480.5 ppm/°C at the nominal temperature of 36.4 °C. Fig. 4.7 illustrates the simulated temperature dependence of I_{SOURCE} at a constant supply voltage of 3.1 V in an (arbitrarily-chosen) temperature interval between 31 °C and 41 °C. The relative error of I_{SOURCE} is smaller than ±0.2 % over the complete temperature interval.

Other errors related to the temperature are:

- The temperature-induced apparent strain due to the temperature dependence of the strain-gauge resistance $R_0(S_i)$. Eq. (3.21) gives the equation for the apparent strain valid for titanium surfaces. In the temperature interval between 31 °C and 41 °C the temperature-related error is smaller than ±5.3 μstrain in comparison with its reference value at 36.4 °C.

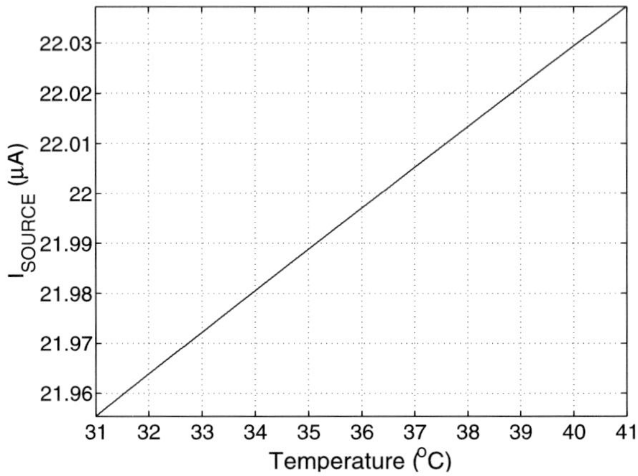

Figure 4.7: *Temperature dependence of I_{SOURCE} at a supply voltage of 3.1 V.*

This is equivalent with a maximum strain-gauge resistance-change $\Delta R_0(S_i)(\Delta T)$ smaller than ±53 mΩ.

- The resistance change of R_{REF} as a function of temperature. R_{REF} is an external ultra-precision chip resistor [Alp] with a nominal resistance of 9310 Ω, an accuracy of ±0.05 % and a temperature coefficient of resistance $TC_{R_{REF}}$ of 0±5 ppm/°C. The *worst-case* error caused by the temperature dependence of R_{REF} over the complete temperature interval between 31 °C and 41 °C is equivalent with a strain of ±13.2 μstrain. The effective error due to R_{REF} though is expected to be smaller because of the mean TCR of 0 ppm/°C. This is the reason why an external resistor has been selected.

- Also the resistance of the wires connecting the strain gauges to the datalogger is dependent on the temperature. If copper (ρ = 16.7x10^{-9} Ωm and TCR = 3.9x10^{-3}/K [Pol 94]) wires are employed with a diameter of 100 μm and a maximum length of 10 cm the maximum temperature-induced error due to both wires connecting the strain gauge S_i is limited to ±0.9 μstrain. The resistance of a single wire R_{wire} at 36.4 °C equals 226 mΩ.

Because I_{REF}, I_{SG} and I_{DAC} are derived from I_{SOURCE}, they all have a temperature dependence $TC_{I_{SOURCE}}$ equal to 480.5 ppm/°C. The gauge factor G of the strain gauges S_i is also temperature dependent (Table 3.2). Its temperature dependence TC_G equals -0.011 %/°C. From Eq. (4.5) follows that the resulting voltage difference as a function of a strain ε and a temperature difference ΔT(=T-36.4 °C) is given by

$$\begin{aligned}\Delta V(\Delta T) = {} & (I_{SG} + I_{DAC})(1 + TC_{I_{SOURCE}} \cdot \Delta T) G (1 + TC_G \cdot \Delta T) \varepsilon R_0'(S_i) \\ & + \left[(I_{SG} + I_{DAC}) R_0''(S_i) - I_{REF} R_{REF} (1 + TC_{R_{REF}} \cdot \Delta T) \right] (1 + TC_{I_{SOURCE}} \cdot \Delta T)\end{aligned} \quad (4.18)$$

where $R_0'(S_i)$=$R_0(S_i)$+$\Delta R_0(S_i)(\Delta T)$ and $R_0''(S_i)$=$R_0'(S_i)$+2$R_{wire}(\Delta T)$. The first term in Eq. (4.18) is proportional to the strain ε to be measured. Because of the negative temperature coefficient

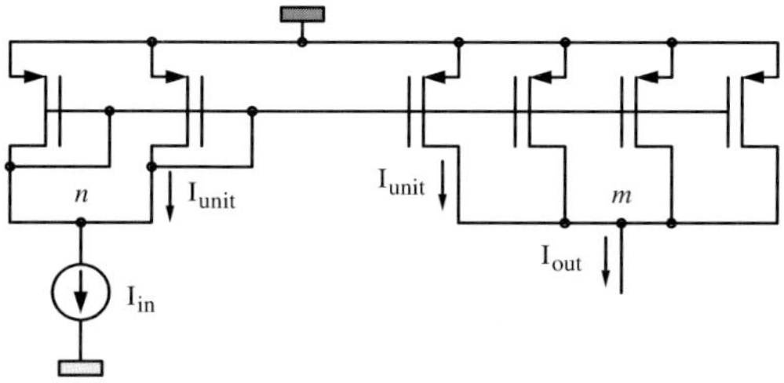

Figure 4.8: *Current mirror with a $\frac{m}{n}$ current ratio and a unit-transistor current mismatch* $\sigma\left(\frac{\Delta I_{unit}}{I_{unit}}\right)$.

of G and the positive temperature coefficient $TC_{I_{SOURCE}}$, the resulting temperature dependence equals approximately 370.5 ppm/°C (neglecting the error due to $R'_0(S_i)$). The maximum error occurs for the maximum/minimum strain $\varepsilon_{max,min}$ and equals approximately ± 3.6 μstrain. The second term in Eq. (4.18) is the residual offset (cf Eq. (4.29)) after compensation. The maximum error due to this term is limited to approximately ± 0.9 μstrain, so that the worst-case error due to the temperature dependence of I_{SOURCE} equals approximately ± 4.5 μstrain.

The resulting *worst-case* temperature-induced error over the complete (arbitrarily-chosen) temperature-interval between 31 °C and 41 °C, occurring at high strain values, equals approximately ± 23.9 μstrain. The normal oral-temperature-variation is equal to 0.5-0.7 °C [Gan 95]. This variation results in a temperature-induced error which is smaller than ± 1.7 μstrain. Because this error is relatively small compared with the desired measurement accuracy, temperature sensors and interface circuits to compensate for the temperature effects, requiring extra space and power, have not been implemented in the datalogger.

4.3.1.5 Current mirror inaccuracy

In addition to the offset due to the strain-gauge nominal-resistance tolerance and pre-strain, the DAC must also compensate for the offset, caused by mismatch in the current mirrors deriving I_{REF}, I_{SG} and I_{DAC} itself from the reference current source I_{SOURCE}. The current mirror, depicted in Fig. 4.8, is used to calculate the output-current inaccuracy resulting from mismatch. It consists of an input transistor M_{in} composed of n unit transistors, and an output transistor M_{out} composed of m unit transistors. The variance of the current mismatch for two unit transistors is given by [Bas 98]

$$\sigma^2\left(\frac{\Delta I_{unit}}{I_{unit}}\right) = \sigma^2\left(\frac{\Delta\beta_{unit}}{\beta_{unit}}\right) + \frac{4\sigma^2(\Delta V_{T,unit})}{(V_{GS}-V_T)^2} \tag{4.19}$$

For the current mirror of Fig. 4.8 the variance of the output-current error is given by (cf Eq. (3.22))

$$\sigma^2\left(\frac{\Delta I_{out}}{I_{out}}\right) = \sigma^2\left(\frac{\Delta\beta_{out}}{\beta_{out}}\right) + \frac{4\sigma^2(\Delta V_{T,out})}{(V_{GS}-V_T)^2} + \frac{4\sigma^2(\Delta V_{GS})}{(V_{GS}-V_T)^2} \tag{4.20}$$

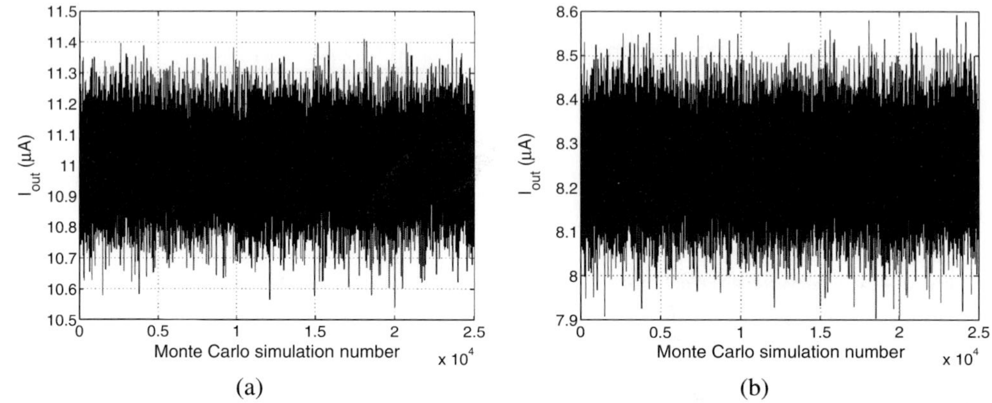

Figure 4.9: *Monte Carlo simulations of the output current I_{out} for two different test cases.*

Note the introduction of an extra term due to the gate-source voltage inaccuracy $\sigma(\Delta V_{GS})$ which is related to the input-transistor mismatch and the input-current inaccuracy. The output transistor consists of m unit transistors that are statistically independent, so that [Bas 98]

$$\sigma^2\left(\frac{\Delta\beta_{out}}{\beta_{out}}\right) = \frac{1}{m}\sigma^2\left(\frac{\Delta\beta_{unit}}{\beta_{unit}}\right) \tag{4.21}$$

and

$$\sigma^2(\Delta V_{T,out}) = \frac{1}{m}\sigma^2(\Delta V_{T,unit}) \tag{4.22}$$

Starting from

$$V_{GS} \approx \sqrt{\frac{I_{in}}{\beta_{in}}} + V_{T,in} \tag{4.23}$$

the variance of the gate-source voltage error can be calculated:

$$\sigma^2(\Delta V_{GS}) = \frac{1}{4}\frac{I_{in}}{\beta_{in}}\sigma^2\left(\frac{\Delta\beta_{in}}{\beta_{in}}\right) + \sigma^2(\Delta V_{T,in}) + \frac{1}{4}\frac{I_{in}}{\beta_{in}}\sigma^2\left(\frac{\Delta I_{in}}{I_{in}}\right) \tag{4.24}$$

In addition to the errors caused by mismatch related to M_{in}, an input-current inaccuracy equal to $\sigma(\Delta I_{in}/I_{in})$ also gives rise to an error, given by the last term in Eq. (4.24). Note that Eq. (4.21) and Eq. (4.22) are applicable for $\sigma^2(\Delta\beta_{in}/\beta_{in})$ and $\sigma^2(\Delta V_{T,in})$ when m is replaced by n. Combination of Eq. (4.21)–Eq. (4.24) results in

$$\sigma^2(\Delta V_{GS}) = \frac{1}{4}(V_{GS} - V_T)^2\frac{1}{n}\sigma^2\left(\frac{\Delta\beta_{unit}}{\beta_{unit}}\right) + \frac{1}{4}(V_{GS} - V_T)^2\sigma^2\left(\frac{\Delta I_{in}}{I_{in}}\right) + \frac{\sigma^2(\Delta V_{T,unit})}{n} \tag{4.25}$$

Simulation	m	n	$\overline{I_{out}}$	$\sigma^2_{calc}(\Delta I_{out}/I_{out})$	$\sigma^2_{sim}(\Delta I_{out}/I_{out})$
Fig. 4.9 (a)	8	4	11 μA	1.0946e-04	1.0906e-04
Fig. 4.9 (b)	6	2	8.2499 μA	1.1681e-04	1.1624e-04

Table 4.2: Comparison between the calculated variance $\sigma^2_{calc}(\Delta I_{out}/I_{out})$ and the simulated variance $\sigma^2_{sim}(\Delta I_{out}/I_{out})$ for the two cases of Fig. 4.9. $\overline{I_{out}}$ is the mean simulated output-current.

By combining this equation and Eq. (4.19)–Eq. (4.22) the following relation is found between the overall inaccuracy of the current mirror, and the unit-transistor mismatch $\sigma(\Delta I_{unit}/I_{unit})$ and the input-current inaccuracy $\sigma(\Delta I_{in}/I_{in})$:

$$\sigma^2\left(\frac{\Delta I_{out}}{I_{out}}\right) = \left(\frac{1}{n}+\frac{1}{m}\right)\sigma^2\left(\frac{\Delta I_{unit}}{I_{unit}}\right) + \sigma^2\left(\frac{\Delta I_{in}}{I_{in}}\right) \tag{4.26}$$

Monte Carlo simulations have been performed to check the validity of this equation. Table 4.2 gives a comparison between the calculated and simulated variance, respectively $\sigma^2_{calc}(\Delta I_{out}/I_{out})$ and $\sigma^2_{sim}(\Delta I_{out}/I_{out})$, for the two examples shown in Fig. 4.9. For both cases the input current I_{in} equals n x 1.375 μA with an accuracy σ $(\Delta I_{in}/I_{in})$ of 1 %. The unit-transistor mismatch $\sigma(\Delta I_{unit}/I_{unit})$ in both cases is equal to 0.5022 %. Table 4.2 demonstrates that the calculated variance and the simulated variance correspond well.

4.3.2 DAC

4.3.2.1 DAC requirements

To compensate for the offsets between V_{S_i} and V_{REF} after placement of the prosthesis, the digitally-controllable current source I_{DAC} is included (Fig. 4.2). In fact, I_{DAC} is a current-steering digital-to-analog converter (DAC), generating an analog current proportional to the digital code applied at its input. The output current I_{DAC} for a N-bit resolution current-steering DAC is given by

$$I_{DAC} = 2^{N-1}d_{N-1}I_{DAC,unit} + ... + 2^1 d_1 I_{DAC,unit} + 2^0 d_0 I_{DAC,unit} = \sum_{i=0}^{N-1} 2^i d_i I_{DAC,unit} \tag{4.27}$$

with $d_{N-1}d_{N-2}...d_2d_1d_0$ the digital input word, d_0 the least significant bit (LSB), d_{N-1} the most significant bit (MSB) and $I_{DAC,unit}$ the DAC unit-current corresponding with 1 LSB.

The *range* of the DAC is determined by the size of the offsets introduced in the two branches of the current-driven Wheatstone configuration. On the one hand the accuracy of V_{REF} is determined by the accuracy of R_{REF} and of I_{REF}, which is dependent on the accuracy of I_{SOURCE} and the current mirror to derive I_{REF} from I_{SOURCE}. On the other hand the accuracy of V_{S_i} is determined by the accuracy of I_{SG}, the tolerance on $R_0(S_i)$, the error introduced by the wires, the error introduced by bonding the gauge to the abutment surface, the pre-strain and the tolerance

Parameter	**Relative error**
Accuracy I_{SOURCE}	$\pm$ 1.5 %
Accuracy R_{REF}	$\pm$ 0.05 %
Accuracy of I_{REF}-current mirror (3σ-approach)	$\pm$ 0.795 %
Tolerance $R_0(S_i)$	$\pm$ 0.6 %
$\Delta R/R_0(S_i)$ due to wires (overestimated)	$\pm$ 0.1 %
$\Delta R/R_0(S_i)$ due to bonding (rule of thumb)	$\pm$ 0.6 %
Tolerance G	$\pm$ 0.5 %
$\Delta R/R_0(S_i)$ due to pre-strain (for G=G_{max})	$\pm$ 0.7 %
Accuracy of I_{SG}-current mirror (3σ-approach)	$\pm$ 0.776 %

Table 4.3: Overview of the errors introducing offsets.

on G. An overview of these errors is given in Table 4.3. The accuracy of the current mirrors to derive I_{REF} and I_{SG} from I_{SOURCE} are calculated with Eq. (4.26) where n=1, m=M (cf Table 4.1) and $\sigma(\Delta I_{unit}/I_{unit})$=0.25 %. The (overestimated) combined worst-case relative error of I_{SOURCE} due to temperature, battery supply-voltage, mismatch and the accuracy of the laser-trimmed resistor R_{PTAT} equals $\pm$ 1.5 %. Note that the sign of the error due to I_{SOURCE} is the same for both branches, since I_{REF} and I_{SG} are both derived from I_{SOURCE}. Note also that the lower tolerance on R and G for metal film strain gauges in comparison with semiconductor strain gauges (Table 3.2 and Table 3.3) results in a smaller compensation range. Eq. (3.13) with G=$G_{max}\approx$2.02 is used to calculate the error due to the pre-strain ($\pm$3232 μstrain). The equivalent accuracy $\Delta R'/R_0(S_i)$ of $R_0(S_i)$, combining the errors related to the strain-gauge resistance, is $\pm$ 2 %.

Combination of the errors results in a worst-case maximum V_{REF} value, denoted as $V_{REF,Max}$, equal to 1.677 V (i.e. for $\Delta I_{SOURCE}/I_{SOURCE}$ = +1.5 %) and a corresponding worst-case minimum value of V_{S_i} equal to 1.520 V. The difference between these two values, i.e. 157 mV, is the worst-case difference that can occur. This is the *range* $\Delta V_{DAC,range}$ over which the DAC must be able to compensate. Note that the values of R_{REF}, I_{REF} and I_{SG} are chosen such that the worst-case maximum value of V_{S_i} (1.606 V) is lower than the corresponding worst-case minimum value of V_{REF} (1.649 V), so that compensation can be achieved by the DAC.

The required *accuracy* and *resolution* of the DAC are determined by the maximum tolerable residual-offset-voltage between V_{S_i} and V_{REF} after compensation. For a perfect compensation the output voltage of the amplifier, explained in Section 4.3.4, following the multi-gauge compensation block equals V_{MM} (1.55 V), i.e. the voltage of the middle terminal of the series-connected batteries. Amplification of the residual input offset-voltage in combination with the input signal due to the maximum/minimum occurring strain $\varepsilon_{max,min}$ must be smaller than the output range of the amplifier in order to avoid overloading of the amplifier. The output range of the amplifier equals 1.1 V (between 1.05 V and 2.15 V) taking into account an ample safety margin. As illustrated in Fig. 4.10, a 0.1 V interval is foreseen for the amplified residual-offset-voltage or in other words the residual input offset-voltage must result in an amplified output voltage between V_{MM} (no residual offset) and 1.65 V. This means that an interval of 0.5 V remains for both

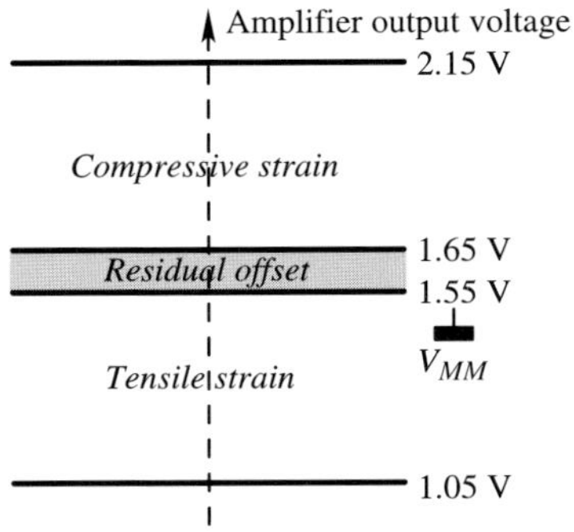

Figure 4.10: *Amplifier output voltage range. A 0.1 V interval is foreseen for the amplified residual offset.*

the compressive and tensile strain measurements (cf Fig. 4.10). The maximum allowable gain $A_{AMP,max}$ of the amplifier without risk of overloading is determined by the maximum/minimum strain $\varepsilon_{max,min}$, resulting in a maximum input-voltage-difference $\Delta V_{in,max}$ when V_{REF} equals $V_{REF,Max}$ (Eq. (4.6)):

$$A_{AMP,max} = \frac{0.5\ V}{\Delta V_{in,max}} = \frac{0.5\ V}{G_{max}\varepsilon_{max,min}V_{REF,Max}} = 75.8 \tag{4.28}$$

The gain of the implemented amplifier A_{AMP} is chosen equal to 70. To ensure that the amplified residual-offset-voltage is smaller than 0.1 V, the maximum voltage step $\Delta V_{DAC,max}$ between two consecutive codes of the DAC is equal to

$$\Delta V_{DAC,max} = \frac{0.1\ V}{A_{AMP}} = 1.43\ mV \tag{4.29}$$

For an ideal DAC with no mismatch between the different current sources the voltage step ΔV_{DAC} between two consecutive DAC codes equals

$$\Delta V_{DAC} = R_0(S_i)I_{DAC,unit} \tag{4.30}$$

ΔV_{DAC} is dependent on the equivalent accuracy $\Delta R'/R_0(S_i)$ ($\pm$ 2 %) of $R_0(S_i)$ and of the accuracy of $I_{DAC,unit}$, which is dependent on the accuracy of I_{SOURCE} ($\pm$ 1.5 %) and the current mirrors to derive $I_{DAC,unit}$ from I_{SOURCE}. The accuracy of $I_{DAC,unit}$ relative to I_{SOURCE} due to the current mirrors equals $\pm$ 1.7 % (3σ-approach) for the chosen architecture and $I_{DAC,unit}$ value as explained further. To derive the required *accuracy* and *resolution* of the DAC an equivalent worst-case maximum and minimum strain-gauge-resistance $R_{0,eq,Max/Min}(S_i)$, combining the different errors, is defined:

$$R_{0,eq,Max/Min}(S_i) = R_0(S_i)(1 \pm 2\%)(1 \pm 1.5\%)(1 \pm 1.7\%) \cong 5000 \pm 265\ \Omega \tag{4.31}$$

The maximum allowable voltage step $\Delta V_{DAC,max}$ and the maximum equivalent strain-gauge-resistance $R_{0,eq,Max}(S_i)$ determine the upper limit of $I_{DAC,unit}$:

$$I_{DAC,unit} < \frac{\Delta V_{DAC,max}}{R_{0,eq,Max}} = \frac{1.43 \text{ mV}}{5265 \ \Omega} = 271.6 \text{ nA} \tag{4.32}$$

$I_{DAC,unit}$ has been chosen equal to 172 nA (= 1 LSB) in the implemented DAC to cope with non-linearity errors introduced by mismatch. In this application the differential non-linearity (DNL) specification of the DAC is the most important accuracy requirement. The DNL-error of a DAC is defined as the maximum deviation of the output step, corresponding to two consecutive codes, from the ideal value of 1 LSB [Raz 95, Joh 97]. This means that in this case the DNL-error must be smaller than +0.58 LSB to comply with Eq. (4.32), which corresponds with the worst-case maximum-allowable current value. The INL-specification, defined as the maximum deviation of the DAC input-output characteristic from the ideal straight-line characteristic [Raz 95, Joh 97], is of lesser importance in this application.

The required *resolution* of the DAC on the other hand is determined by the minimum equivalent strain-gauge-resistance $R_{0,eq,Min}$. The minimum required number of steps is determined by the minimum mean voltage step $\Delta V_{DAC,min}$, corresponding with $R_{0,eq,Min}$, and the range of the DAC $\Delta V_{DAC,range}$:

$$\text{number of steps} > \frac{\Delta V_{DAC,range}}{\Delta V_{DAC,min}} = \frac{\Delta V_{DAC,range}}{R_{0,eq,Min} I_{DAC,unit}} = \frac{157 \text{ mV}}{814.4 \ \mu\text{V}} = 193 \tag{4.33}$$

which means that the required number of bits or *resolution* of the DAC equals 8.

4.3.2.2 Operating principle and implementation

Despite the fact that a unit element architecture has a better performance concerning the DNL-specification, a binary-weighted architecture has been selected for the current-steering DAC, because of its minimum decoding logic complexity and area [Bas 98]. A simplified schematic of the implemented 8-bit binary-weighted architecture is shown in Fig. 4.11. It consists of 8 current sources providing a current equal to $2^i I_{DAC,unit}$ ($i = 0..7$). Each of these current sources is composed of 2^i unit-current-source PMOS transistors M_{cur}, also called unit current sources in the remainder of the text. The 8 switches are controlled by the digital input word $d_7 d_6 ... d_2 d_1 d_0$ and the output current I_{DAC} is given by Eq. (4.27).

The worst-case DNL-error of the implemented DAC must be smaller than +0.58 LSB. To satisfy this condition the DAC is designed with an INL-specification of 0.23 LSB. If this INL-specification is met, then the *maximum* DNL-error of the DAC is smaller than 0.46 LSB [Bas 98]. A margin of 0.12 LSB remains for the systematic DNL-errors discussed further. The DAC yield , defined as the percentage of functional devices that have an INL-error less than or equal to the required INL-specification, is dependent on the current mismatch σ_I/I of the unit current sources. Because the DNL-specification is the most important non-linearity specification in this case and because statistically in a binary implementation the maximum DNL-error has the highest probability of occurring at the MSB transition [Bas 98], the MSB-transition yield-model of

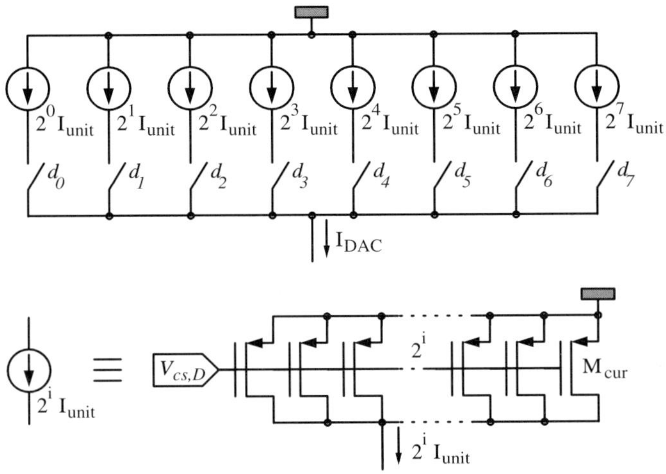

Figure 4.11: *Binary-weighted architecture DAC.* $I_{unit}=I_{DAC,unit}$.

Lakshimikumar [Lak 88] is used to calculate the required σ_I/I. For a 8-bit resolution DAC with a 0.23-LSB INL-specification the yield Y, predicted by this model, is given by [Lak 88]

$$Y = \prod_{i=127}^{128} \operatorname{erf}\left(\frac{Q_i}{\sqrt{2}}\right) \tag{4.34}$$

with

$$Q_i = \frac{0.23}{256\left[\frac{\overline{z_i}(1-\overline{z_i})}{255}\right]^{1/2}\left(\frac{\sigma_I}{I}\right)} \tag{4.35}$$

and

$$\overline{z_i} = \frac{i}{255} \tag{4.36}$$

Fig. 4.12 shows the yield predicted by Eq. (4.34). In order to achieve a yield better than 99 % the current mismatch σ_I/I must be smaller than 1.02 %. σ_I/I is dependent on the size of the unit current sources and on their overdrive voltage V_{GST}, i.e. V_{GS}-V_T. The minimum unit-current-source area required to achieve a current mismatch σ_I/I is given by [Bas 98]

$$W_{cur}L_{cur} \geq \frac{1}{2}\left[A_\beta^2 + \frac{4A_{VT}^2}{(V_{GS}-V_T)^2}\right] / \left(\frac{\sigma_I}{I}\right)^2 \tag{4.37}$$

with W_{cur} and L_{cur} the width and length of M_{cur}, and A_β and A_{VT} the mismatch proportionality parameters. A_β and A_{VT} are equal to 2.8 %μm and 22 mVμm respectively for the Alcatel Microelectronics C07MA technology [Bas 98]. The unit current sources are implemented with

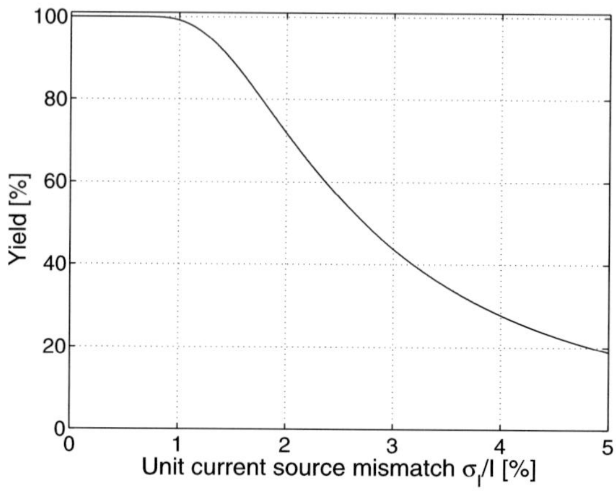

Figure 4.12: *DAC yield for a 8-bit resolution DAC with a 0.23-LSB INL-specification as a function of* σ_I/I.

a V_{GST} equal to -0.5 V. From Eq. (4.37) follows then that for a current mismatch σ_I/I of 1.02 % the minimum required area $W_{cur}L_{cur}$ equals 41 μm^2. On the other hand the W_{cur}/L_{cur} ratio determines the current of the unit current sources, which must be equal to $I_{DAC,unit}$. These two conditions for W_{cur} and L_{cur} result in

$$W_{cur} = 1.33\ \mu m \text{ and } L_{cur} = 31\ \mu m \tag{4.38}$$

In order not to waste any silicon area because of the routing of the metal lines to connect the unit current sources [Bas 98], which are implemented in a current source array, and to obtain a better current-source aspect-ratio, i.e. closer to 1, a new unit current source $M_{cur,new}$ is defined. The new unit current source is equivalent with 8 old unit current sources. Its current $I_{DAC,unit,new}$ equals 1375 nA and its dimensions are given by

$$W_{cur,new} = 10.6\ \mu m \text{ and } L_{cur,new} = 31\ \mu m \tag{4.39}$$

As shown in Fig. 4.13 the 4 MSB current sources of the 'new' implemented DAC are realized with parallel-connected new unit current sources, while the 3 LSB current sources are realized with series-connected new unit current sources.

A drawback of this approach is that the current divisions, realized by the series-connected current sources, are not perfect. Table 4.4 gives an overview of the simulated currents of the 3 LSB current sources. To check whether the DNL-specification is not jeopardized by this approach, the systematic DNL-error resulting from the non-perfect currents has been simulated with MATLAB. Fig. 4.14 illustrates the result of this simulation and shows that the resulting systematic DNL-error has a maximum value of +0.06 LSB and a minimum value of -0.3 LSB. In this application only positive DNL-errors are critical, because negative errors correspond with

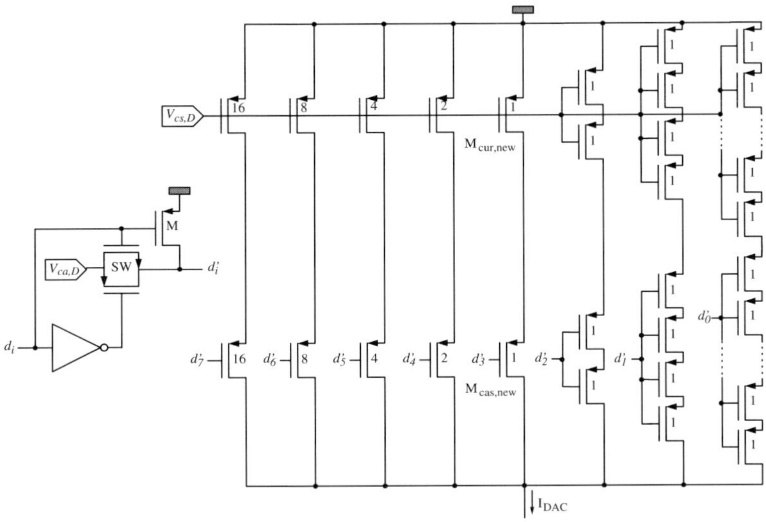

Figure 4.13: *Final DAC-implementation. Transistor dimensions are listed in Table A.2.*

steps smaller than 1 LSB, which do not cause a problem to carry out the wanted compensation. The *maximum* positive DNL-error due to mismatch and the non-perfect divisions equals +0.52 LSB.

The 5 MSB bits of the DAC (cf Fig. 4.13) are composed of 31 parallel-connected new unit current sources, which are implemented in a 6x6 current-source-array. In order to reduce the systematic errors, introduced by errors in the current source array, which are to first order linear (graded) and quadratic (symmetric) in spatial distribution [VdP 01, Vdb 02], a common-centroid layout is applied for the array [San 94]. The current sources of the LSB bits are placed around this array in two additional rows and columns and provide identical surroundings to the inner array, minimizing edge effects. To provide also identical surroundings to the LSB-bits current-sources two more rows and columns with dummy current sources are added, so that the final implemented current source array incorporates 10 rows and 10 columns.

The 3 current sources of the 3 LSB bits, as well as one of the current sources of the 6x6

Bit	Ideal current	Simulated current	Error
0	1375/8 nA	181.95 nA	+0.0586 LSB
1	1375/4 nA	361.35 nA	+0.1024 LSB
2	1375/2 nA	711.48 nA	+0.1395 LSB

Table 4.4: Overview of the 3 LSB currents, realized by series-connected current sources.

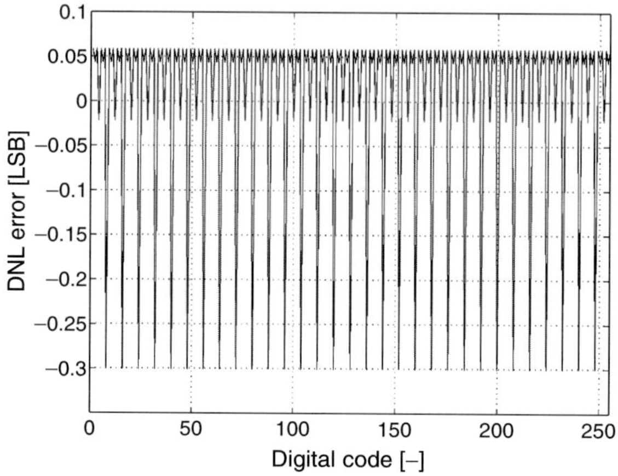

Figure 4.14: *Systematic DNL-error introduced by the non-perfect current divisions.*

current-source-array, corresponding with bit d_3, are left uncompensated by the common-centroid layout. This gives rise to systematic errors, resulting from graded process variations in the current source array. The total weight of the uncompensated current sources equals 15, so that the upper bound $\mathrm{DNL_{sys,Max}}$ for the resulting systematic DNL-error is equal to [Bas 98]

$$\mathrm{DNL_{sys,Max}} \leq 2{\cdot}\mathrm{INL_{sys,Max}} = 2{\cdot}15{\cdot}\frac{E_{p-p}}{2} \tag{4.40}$$

with E_{p-p} the worst-case peak-to-peak error given by

$$E_{p-p} = D(S_\beta + \frac{2}{\mathrm{V_{GST}}}S_{VT})\ \mathrm{LSB} \tag{4.41}$$

with $S_\beta = 0.3$ %/mm and $S_{VT} = 0.1$ mV/mm the worst-case process proportionality parameters and $D = 420\ \mu$m the maximum separation distance between two current sources of the DAC. In fact, the worst-case error $\mathrm{DNL_{sys,Max}}$ is included here only for safety, because it is demonstrated in [Bas 98] that for the C07MA technology the mismatch degradation as a function of distance can be ignored for separation distances smaller than 500 μm. The symmetric error, which would result from eutectic die-bonding [Bas 98], is not taken into account in the above calculations, because this bonding technique is not used here. Eq. (4.40) results in a maximum systematic DNL-error (3σ-approach) of 0.06 LSB, so that the total *maximum* positive DNL-error equals +0.58 LSB and the DNL-specification is satisfied.

The current sources $\mathrm{M_{cas,new}}$ of the final DAC-implementation, shown in Fig. 4.13, are biased by the current-source voltage $V_{cs,D}$. To reduce the sensitivity of $\mathrm{I_{DAC}}$ to the DAC's output voltage, (wide-swing) cascode transistors $\mathrm{M_{cas,new}}$ are implemented. The 8 cascode transistors scale with the 8 current sources in order to obtain equal 'drain-source'-voltages for the latter. The cascode transistors, controlled by d_i', also function as switches. When d_i is high, the gate voltage

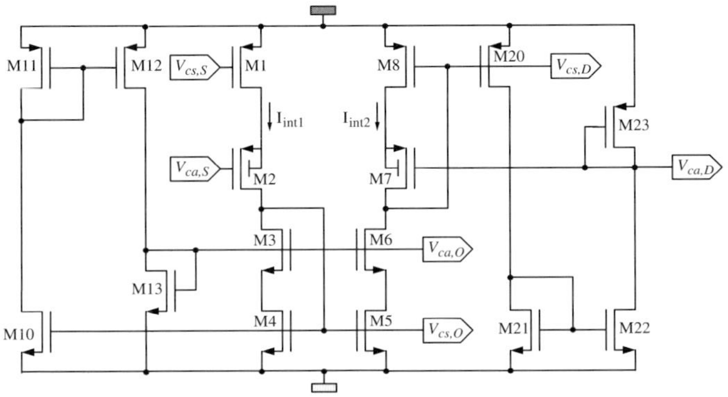

V_{DD}	3.1 V
V_{SS}	0 V
$V_{cs,S}$	1.948 V
$V_{ca,S}$	1.602 V
$V_{cs,O}$	1.426 V
$V_{ca,O}$	1.980 V
$V_{cs,D}$	1.836 V
$V_{ca,D}$	1.510 V

Figure 4.15: *Derivation of $V_{cs,D}$ and $V_{ca,D}$ from the reference current source I_{SOURCE} (cf Table A.3 for transistor dimensions).*

d_i' of the cascode is connected to the cascode voltage $V_{ca,D}$ and the cascode conducts the current of the corresponding current source. When d_i is low, d_i' is connected to the positive supply voltage V_{DD}, switching off the cascode. In order to reduce the power consumption, global switches are implemented to switch on/off the combined current I_{SG}+I_{DAC} through the strain gauges S_i. This will be explained in Section 4.3.4.2.

4.3.2.3 Derivation and accuracy of the new unit current source

The new unit current sources $I_{DAC,unit,new}$ are obtained from I_{SOURCE} by means of 3 wide-swing cascode current mirrors. Fig. 4.15 shows how the current-source voltage $V_{cs,D}$ and the cascode voltage $V_{ca,D}$ (Fig. 4.13) are derived from I_{SOURCE}. $V_{cs,S}$ and $V_{ca,S}$ are the current-source voltage and the cascode voltage of the wide-swing cascode current mirror of the reference current source, as illustrated in Fig. 4.4. $V_{cs,O}$ and $V_{ca,O}$ are used to bias the OTA's of the succeeding amplifier and sample-and-hold (cf Fig. 4.19).

Table 4.5 gives an overview of the accuracy of the current mirrors employed to derive the current source, corresponding with bit d_3, from I_{SOURCE}. This current source consists of a single unit current source $I_{DAC,unit,new}$. It can be seen from Fig. 4.13 that the worst-case accuracy applies for this current source, because the number of transistors (i.e. m) in the output branch

Current mirror	n	m	$\sigma^2(\Delta I_{in}/I_{in})$	$\sigma(\Delta I_{unit}/I_{unit})$	$\sigma^2(\Delta I_{out}/I_{out})$
I_{SOURCE}-I_{int1}	1	1	0	0.25 %	.125e-4
I_{int1}-I_{int2}	5	5	.125e-4	0.25 %	.15e-4
I_{int2}-$I_{DAC,unit,new}$	16	1	.15e-4	0.4 %	.32e-4

Table 4.5: Overview of the accuracy of the current mirrors deriving $I_{DAC,unit,new}$ from I_{SOURCE}.

of the current mirror used to derive this current source is equal to 1, resulting in a high relative output-current inaccuracy according to Eq. (4.26). The calculated worst-case relative accuracy of $\pm$ 1.7 % (3σ-approach) has been used for $I_{DAC,unit}$ in the calculations above. This error is related to the accuracy of the slope of the input-output characteristic of the DAC or in other words related to the gain error of the DAC.

4.3.3 PROG/SEL-block

4.3.3.1 Implementation

After placement of the prosthesis the DAC input words $d_7d_6...d_2d_1d_0$, required for compensation of the strain-gauge channels, are programmed into the nulling memory REG by means of the PROG/SEL-block (Fig. 4.3). The nulling memory REG consists of 8-bit registers, each corresponding with a different strain-gauge channel. In the measurement mode the compensation word, stored in the nulling memory register associated with the selected channel, is applied to the input of the DAC. Each time a new channel is selected, the compensation word corresponding with that channel is applied to the DAC. In this way offset-compensated measurements are performed. The selection between the different channels is carried out by multiplexers (MUX).

The PROG/SEL-block has two operation modes:

- *Measurement mode*: in this mode the PROG/SEL-block selects a strain-gauge channel dependent on the input word $in_1in_2in_3in_4in_5$ and applies the compensation word related to the selected strain-gauge channel to the input of the DAC.
- *Programmation mode*: in this mode the compensation word required for a particular channel is programmed in the related nulling memory register.

Fig. 4.16 shows the PROG/SEL-block implementation. The selection between the *measurement mode* and the *programmation mode* is done with *prog*. The input word $in_1in_2in_3in_4in_5$ is used to select the wanted strain-gauge channel and to program the nulling memory registers. Also sel_1 and sel_2 are employed for the programmation of the registers as explained further. The channel-number word $c_1c_2c_3c_4c_5$ and inverse channel-number word $\overline{c_1}\,\overline{c_2}\,\overline{c_3}\,\overline{c_4}\,\overline{c_5}$ form the input of a 5-bit binary decoder used to select the wanted strain-gauge channel. The decoder's output channel-selection bit s_i, which corresponds with the decimal equivalent (i.e. i) of $c_1c_2c_3c_4c_5$, is set high, while the other output channel-selection bits are set low. In this way strain-gauge channel i is selected for measurement/programmation. The decoder consists of 5-bit ANDs of which the inputs correspond with the binary codes of the channels (Fig. 4.16). In the *measurement mode* (*prog* low) $c_1c_2c_3c_4c_5$ equals $c''_1c''_2c''_3c''_4c''_5$ which in turn is equal to $in_1in_2in_3in_4in_5$. The strain-gauge channel i corresponding with the decimal equivalent of $in_1in_2in_3in_4in_5$ is selected by the channel-selection bit s_i and the contents of register i of the nulling memory REG is applied to the input of the DAC. In the *programmation mode* (*prog* high) $c_1c_2c_3c_4c_5$ equals $c'_1c'_2c'_3c'_4c'_5$ which in turn is equal to the word stored in the 5 SR (Set-Reset) flip-flops [Rab 96]. The programmation of these flip-flops is done by means of $in_1in_2in_3in_4in_5$, sel_1 and sel_2. When sel_1 and sel_2 are both high and one of the two goes low, the input word $in_1in_2in_3in_4in_5$ is stored

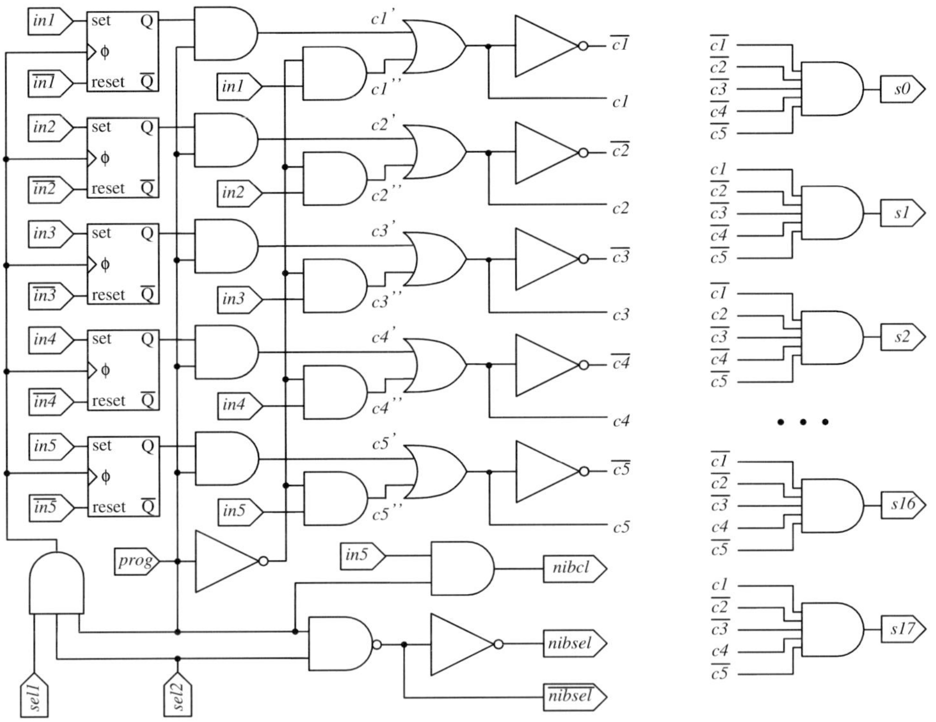

Figure 4.16: *Schematic overview of the PROG/SEL-block.*

into these flip-flops. At the beginning of the programming phase the strain-gauge channel number is stored into the flip-flops to ensure that the same channel and associated register are selected during the whole programmation phase. This is necessary because $in_1in_2in_3in_4in_5$ are also used for the programmation of the MSB and LSB nibbles (i.e. 4 bits) of the register associated with the selected channel. In this way the number of input lines is reduced.

The nulling memory REG consists of programmable 8-bit registers. One of these is shown in Fig. 4.17. It is composed of 8 SR flip-flops followed by tri-state buffers. When the register is selected by the channel-selection bit s_i the register output $d_7d_6...d_2d_1d_0$ is equal to the word stored in the flip-flops. When it is not selected, the output bits $d_7d_6...d_2d_1d_0$ are high-impedant (Z). During programmation of the register sel_2 is used to select between the register's MSB and LSB nibble. When sel_2 is high, *nibsel* is high and the MSB nibble is selected for programmation. On the other hand, when sel_2 is low, the LSB nibble is selected. *nibcl* is used to clock the 8 SR flip-flops. When a high-to-low transition of the LSB input bit in_5 occurs, *nibcl* also goes from high to low and the 4 MSB input bits in_1, in_2, in_3 and in_4 are stored in the selected nibble. In this way both nibbles of the register can be programmed.

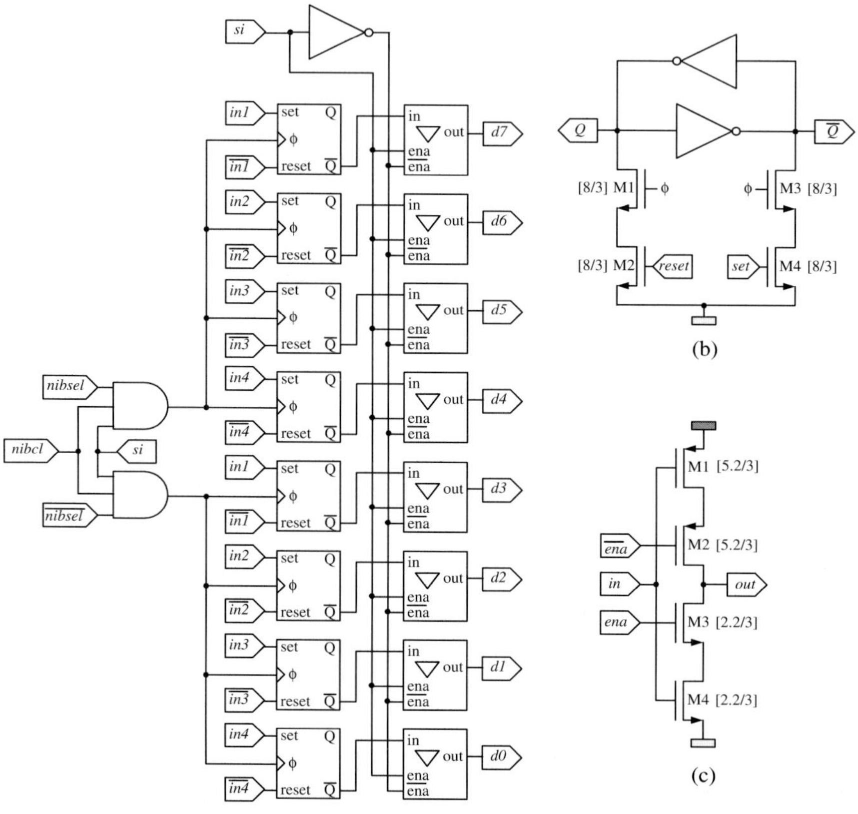

Figure 4.17: *(a) 8-bit register with tri-state outputs. (b) SR flip-flop. (c) Single tri-state buffer.*

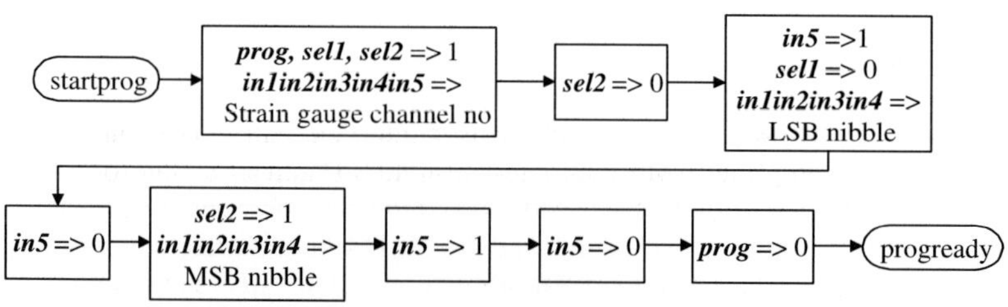

Figure 4.18: *Register programming-protocol.*

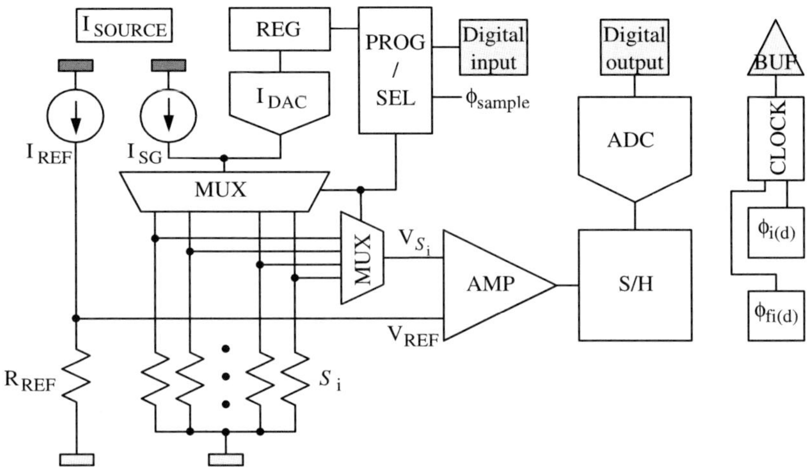

Figure 4.19: *Schematic overview of the complete sensor interface chip.*

4.3.3.2 Programming protocol

Fig. 4.18 illustrates the protocol to program a strain-gauge channel register. First the strain-gauge channel number of the register that needs to be programmed is applied to $in_1in_2in_3in_4in_5$, while $prog$, sel_1 and sel_2 are high. Next sel_2 goes from high to low and the strain-gauge channel number is stored into the flip-flops of the PROG/SEL-block. The register corresponding with the chosen strain-gauge channel is selected for the remainder of the programming phase. Thereafter, in_5 is set high, sel_1 is set low and the LSB nibble data are applied to $in_1in_2in_3in_4$. Next, in_5 goes low and the LSB nibble of the register is stored. Then sel_2 is set high and the MSB nibble is put at the input. Next, a high-to-low transition of in_5 is applied, so that also the MSB nibble is stored. The programming phase ends by setting *prog* low. The above described procedure is implemented in the digital part as explained in Section 5.6.1.

4.3.4 Amplifi er

4.3.4.1 Operating principle

Fig. 4.19 gives an overview of the complete sensor interface chip. The multi-gauge nulling block is followed by an amplifier AMP, a sample-and-hold S/H and an analog-to-digital converter ADC. Also a 128-kHz relaxation clock-oscillator CLOCK and 2 bi-phasic non-overlapping clock generators $\phi_{\{1,2\}(d)}$ ($\phi_{i(d)}$) and $\phi_{f\{1,2\}(d)}$ ($\phi_{fi\,(d)}$) are implemented. The latter have a period of respectively 2 kHz and 64 kHz. They both consist of two different bi-phasic non-overlapping clocks with a slightly different timing: ϕ_i and ϕ_{fi} are slightly advanced in comparison with respectively ϕ_{id} and $\phi_{fi\,d}$. The reason for this will be explained in Section 4.3.4.6.

Because of the low-power requirement of the datalogger the current through the strain gauges

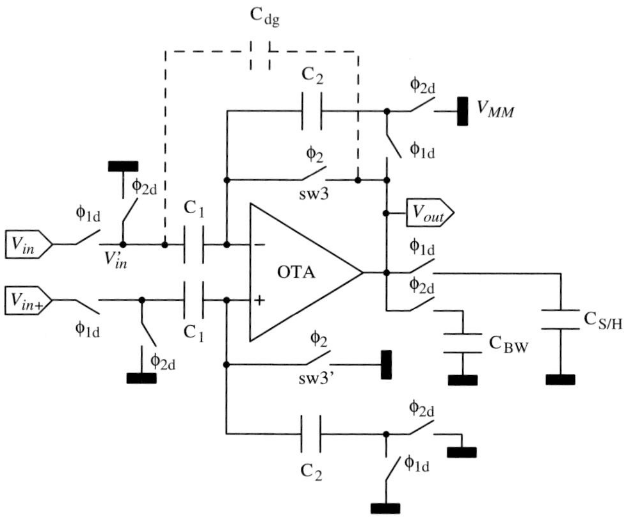

Figure 4.20: *Switched-capacitor amplifier with offset-cancellation.*

S_i is kept to a minimum. This gives rise to small sensor signals. Therefore, the (temperature-dependent) offset of the amplifier, typically of the order of 1-10 mV [Enz 96], and the drift of this offset can cause problems for the sensor interface. To cope with these a switched-capacitor resettable gain amplifier, shown in Fig. 4.20, has been implemented. This amplifier includes offset-cancellation, based on the principle of Correlated Double Sampling (CDS). The basic idea behind this technique is the sampling and storing of the OTA's offset during one phase (reset phase) and 'subtracting' the sampled offset from the offset occurring during the next phase (amplification phase). Because the offset variation with temperature and the drift of the offset are slowly-varying signals, the two offset values are strongly correlated, given that the time between the two phases is sufficiently small. By 'subtraction' of the offset values of the two successive phases offset-cancellation is achieved. The CDS principle is not only used to cancel the OTA's offset, but also to reduce the OTA's 1/f-noise. Because the low-frequent character of this noise, two subsequent 1/f-noise values are strongly correlated too, so that also the 1/f-noise noise contribution is reduced by this technique. The CDS technique can also be applied to lower the sensitivity of the circuit performance to a finite OTA gain as explained in [Enz 96].

Since the CDS technique uses sampling in order to reduce the offset and 1/f noise, the OTA's white noise is aliased by this technique. This does not impose a problem for the system under study, because the multi-gauge sensor interface is inherently a sampled-data system, so that the baseband noise is not deteriorated by this technique.

Another way of looking at the effect of CDS is to note that it is equivalent with subtracting from the occurring time-varying noise a recent sample of the same noise. For DC or very low-frequency noise this results in a cancellation. This indicates that CDS high-pass filters the occurring noise. More details about the exact transfer functions for both the 1/f noise and white

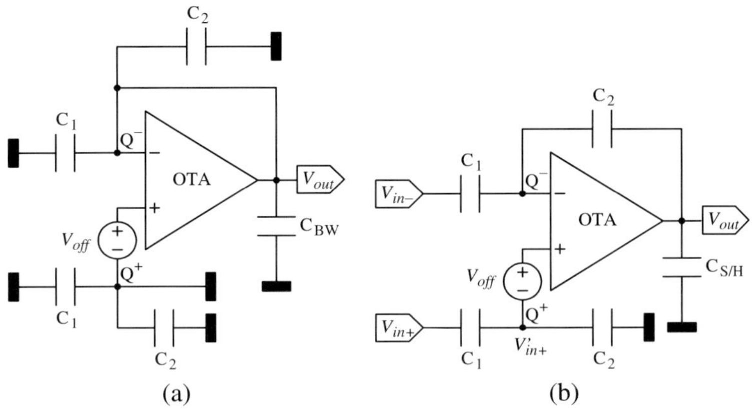

Figure 4.21: *The resettable-gain-amplifier circuit (a) during the reset phase ϕ_2 and (b) during the amplification phase ϕ_1.*

noise can be found in [Enz 96].

To understand the operation principle of the resettable gain amplifier the equivalent circuits in respectively the reset phase ϕ_2 and the amplification phase ϕ_1 are shown in Fig. 4.21. Note that $C_{S/H}$ is the input capacitor of the sample-and-hold S/H (which is assumed ideal) and that the capacitor C_{BW} is shared between S/H and the amplifier. In this way the bandwidth is limited in both their reset phases resulting in smaller noise contributions (cf Section 4.3.4.7). An infinite OTA gain and full settling at the end of the two phases are assumed. To simplify the expressions the voltage V_{MM} is assumed 0 V (i.e. grounded) in the calculations. This only affects the resulting DC-component. The offset voltage V_{off} of the OTA is modeled by a voltage source in series with the positive input of the OTA. The charges at the positive (Q^+) and negative input (Q^-) of the OTA at the end of ϕ_1 are the same as they are at the end of the previous ϕ_2 phase, because there exists no conductive path for these charges to flow during ϕ_1. Therefore, the law of charge conservation can be applied for Q^+ and Q^-

$$Q^+_{\phi_1} = Q^+_{\phi_2} \quad (4.42)$$

$$Q^-_{\phi_1} = Q^-_{\phi_2} \quad (4.43)$$

where the charges at the end of the phases are considered. From Eq. (4.42) and Eq. (4.43) the input-output relation of the resettable gain amplifier can be found. First the end of the reset phase ϕ_2 is considered. The total amount of charge at the negative ($Q^-_{\phi_2}$) and positive ($Q^+_{\phi_2}$) OTA input is given by

$$Q^+_{\phi_2} = 0 \quad (4.44)$$

$$Q^-_{\phi_2} = C_1 \cdot V_{off} + C_2 \cdot V_{off} \quad (4.45)$$

The charges are assumed positive at the OTA inputs. At the end of subsequent amplification

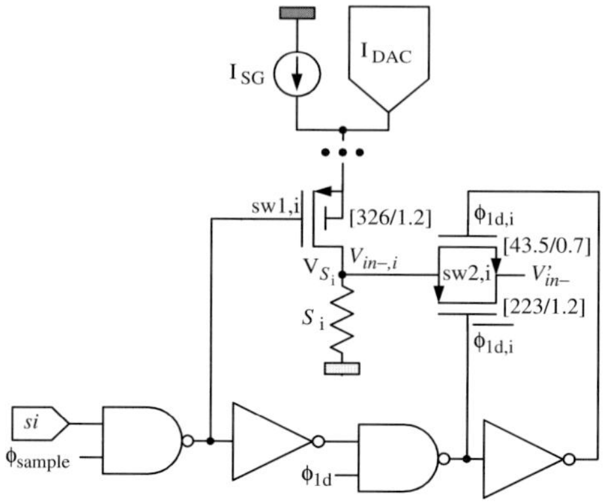

Figure 4.22: *Multiplexers for the selection of the strain-gauge-channel.*

phase ϕ_1 the total amount of charge at the negative and positive input is given by

$$Q^+_{\phi_1} = (V'_{in+} - V_{in+}) \cdot C_1 + V'_{in+} \cdot C_2 \tag{4.46}$$

$$Q^-_{\phi_1} = (V'_{in+} + V_{off} - V_{in-}) \cdot C_1 + (V'_{in+} + V_{off} - V_{out}) \cdot C_2 \tag{4.47}$$

Combining Eq. (4.42), Eq. (4.44) and Eq. (4.46) yields

$$V'_{in+} = \frac{C_1}{C_1 + C_2} \cdot V_{in+} \tag{4.48}$$

and from Eq. (4.43), Eq. (4.45), Eq. (4.47) and Eq. (4.48) follows

$$V_{out} = \frac{C_1}{C_2} \cdot (V_{in+} - V_{in-}) \tag{4.49}$$

It can be seen that the terms containing V_{off} cancel each other. The gain of the resettable gain amplifier is given by the ratio of the two capacitors C_1 and C_2. In order to limit the noise contributions (cf Section 4.3.4.7) C_1 and C_2 have been chosen equal to 770 pF and 11 pF, so that the gain of the implemented amplifier A_{AMP} is equal to 70, satisfying Eq. (4.28) (cf Section 4.3.2.1).

Note that the small (0.5 pF) deglitching capacitor C_{dg} (Fig. 4.20) does not play a role in the signal charge redistribution. Its sole purpose is to prevent glitches in the OTA output by providing negative feedback during the brief intervals when the non-overlapping clock phases are both low, and the feedback path of the OTA would otherwise be open-circuited [Mat 87].

4.3.4.2 MUX

Fig. 4.22 shows the implementation of the multiplexers MUX (cf Fig. 4.3), consisting of the PMOS transistors sw1,i and the transmission gates sw2,i. The sw1,i-switches are used to connect a strain-gauge channel *i* to the current sources I_{SG} and I_{DAC}, while the sw2,i-switches are

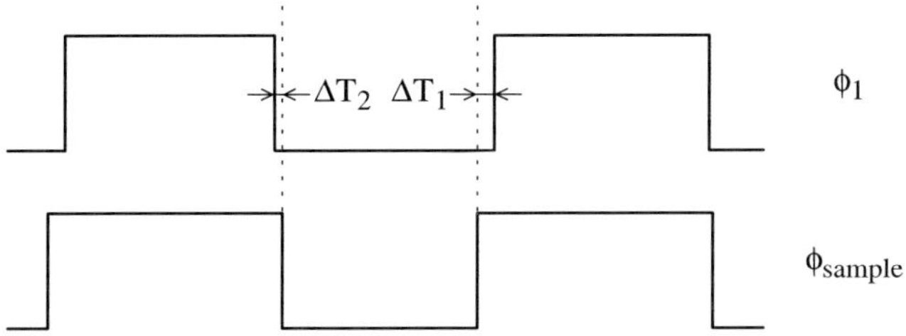

Figure 4.23: *Timing of ϕ_{sample} compared with ϕ_1.*

employed to apply the strain-gauge voltage V_{S_i} to the amplifier AMP. The sw2,i-switches are equivalent with the input switch of the amplifier AMP between V_{in-} (i.e. V_{S_i}) and V'_{in-} (cf Fig. 4.20). The input switch of the AMP is actually composed of 18 different sw2,i-switches. When a strain-gauge channel is selected, the corresponding sw2,i-switch is clocked by ϕ_{1d}.

When a strain-gauge channel *i* is selected by the channel-selection bit s_i, the current I_{SG}+I_{DAC} flows trough S_i when ϕ_{sample} is high. ϕ_{sample} is a special clock derived from ϕ_1 as explained in Section 4.3.7.2. Fig. 4.23 shows the timing of ϕ_{sample} compared with ϕ_1 where $\Delta T_1 \approx$ 5x1/128kHz $\approx$ 39 μs and $\Delta T_2 \approx$ 1x1/128kHz $\approx$ 7.8 μs. The advantages of using the special clock ϕ_{sample} are the restriction of the switching effects at the clock transitions of critical importance for the AMP, and the reduction of the mean power consumption.

ϕ_{sample} has a nominal frequency of 2 kHz. When ϕ_{sample} is low, another channel can be selected by the multiplexers MUX. The 18 channels are sampled one after the other by applying successively the different channel numbers at the input of the sensor interface. This means that each individual channel is sampled at a frequency of 111 Hz.

4.3.4.3 Finite OTA gain

In the previous derivation the gain of the OTA has been assumed infinite. To investigate the influence of the OTA's finite gain, the two phases of the amplifier are reconsidered (Fig. 4.24) and now the OTA is represented by a voltage-controlled voltage-source with gain -A. The equations found above for the charges Q^+ at the positive OTA input are still valid. The equations for the charges at the negative input in the two phases become

$$Q^-_{\phi_2} = C_1 \cdot V_{off} \cdot \frac{A}{1+A} + C_2 \cdot V_{off} \cdot \frac{A}{1+A} \tag{4.50}$$

$$Q^-_{\phi_1} = (V'_{in+} + V_{off} - \frac{V_{out}}{A} - V_{in-}) \cdot C_1 + (V'_{in+} + V_{off} - \frac{V_{out}}{A} - V_{out}) \cdot C_2 \tag{4.51}$$

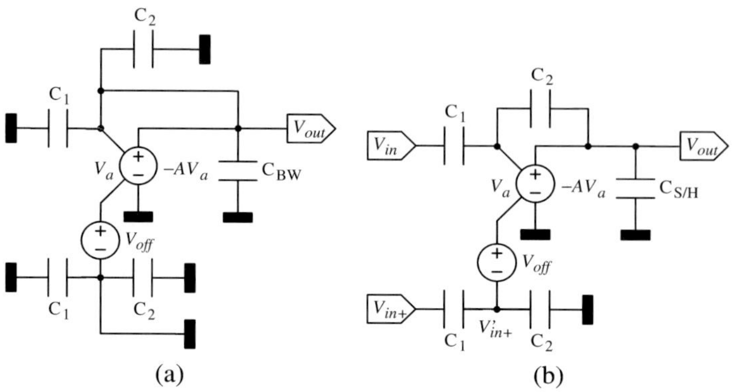

Figure 4.24: *The resettable-gain-amplifier circuit considering an OTA with a finite gain A (a) during the reset phase ϕ_2 and (b) during the amplification phase ϕ_1.*

By applying the law of charge conservation the following input-output relation

$$\begin{aligned} V_{out} &= \frac{C_1}{C_2} \cdot (V_{in+} - V_{in-}) \cdot \frac{1}{1 + \frac{1}{A \cdot f_{dc1}}} + V_{off} \cdot \frac{\frac{1}{1+A} \cdot \frac{1}{f_{dc1}}}{1 + \frac{1}{A \cdot f_{dc1}}} \\ &\approx \frac{C_1}{C_2} \cdot (V_{in+} - V_{in-}) \cdot (1 - \varepsilon_s) + \frac{V_{off}}{A \cdot f_{dc1}} \end{aligned} \tag{4.52}$$

is found with

$$f_{dc1} = \frac{C_2}{C_1 + C_2} \tag{4.53}$$

the feedback factor during the amplification phase. The gain A of the implemented OTA equals approximately 16900, so that the resulting static error ε_s is equal to 0.42 %, which results in a maximum error of ± 8.2 μstrain corresponding with the maximum/minimum strain $\varepsilon_{max,min}$. This however does not impose a problem, since the different strain-gauge channels are calibrated before the actual measurements start. In this way the static error due to the finite OTA gain is calibrated for.

From Eq. (4.52) follows that the resulting offset error in the output is reduced by a factor equal to the product of the gain A and the feedback factor f_{dc1}. The input-referred offset σ_{offset} of the implemented folded-cascode OTA (Fig. 4.30) is given by

$$\sigma^2_{offset} \approx \sigma^2_{M1a/b} + \left(\frac{g_{m7}}{g_{m1}} \cdot \sigma_{M6/7} \right)^2 + \left(\frac{g_{m9}}{g_{m1}} \cdot \sigma_{M8/9} \right)^2 \tag{4.54}$$

where the offset voltage of two matched transistors σ_{Mi} is expressed as a function of their overdrive voltage and gate area by [Pel 89]

$$\sigma^2_{Mi} = \frac{A^2_{VT}}{(WL)_i} + \frac{(V_{GS} - V_T)^2_i}{4} \cdot \frac{A^2_{\beta}}{(WL)_i} \tag{4.55}$$

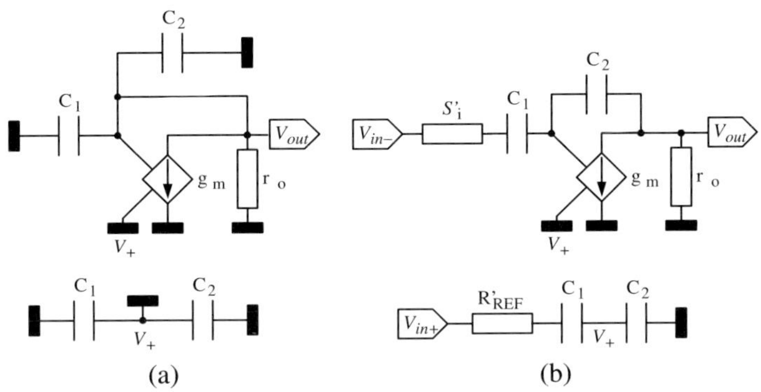

Figure 4.25: *The resettable-gain-amplifier circuit considering an OTA modeled by a transconductance g_m and an output resistance r_o (a) during ϕ_2 and (b) during ϕ_1.*

The calculated input-referred offset voltage σ_{offset} of the implemented OTA equals 0.8 mV. This offset voltage is reduced by the factor $A \cdot f_{dc1}$, so that the final error in the output due to the offset voltage equals approximately 3.4 μV, which can be neglected.

4.3.4.4 Settling behavior

In the previous model no settling effects are taken into account. The voltages are assumed to take their final value instantaneously. The pole(s) of the OTA however limit the settling performance. Therefore the model is changed to the one shown in Fig. 4.25. The OTA is now modeled by a transconductance $\mathrm{g_m}$ and a finite output resistance $\mathrm{r_o}$. The gain A of the OTA is equal to

$$A = -\mathrm{g_m} \cdot \mathrm{r_o} \tag{4.56}$$

The switches are assumed ideal (i.e. with a zero on-resistance) except for the input switches sw2,i (cf Fig. 4.22) at both sides of the OTA. Because of their relevance to the settling behavior, they are combined with S_{i} and $\mathrm{R_{REF}}$:

$$S'_{\mathrm{i}} = S_{\mathrm{i}} + \mathrm{R_1} \tag{4.57}$$

$$\mathrm{R'_{REF}} = \mathrm{R_{REF}} + \mathrm{R'_1} \tag{4.58}$$

where $\mathrm{R_1}$ and $\mathrm{R'_1}$ are the maximum resistance values of the input switches at the negative and positive side respectively. Note that the Norton-Thévenin theorem is applied here to replace the current sources. The new model is simplified in comparison with the previous one as far as the capacitances at the output of the OTA are concerned. This is done to reduce the complexity of the equations in order to improve the understanding. In the second part of the calculations the capacitances at the OTA output are included again as well as the various parasitic capacitances.

In the amplification phase (Fig. 4.25 (b)) the pole of the circuit at the positive OTA input is given by

$$p_{pos} \approx \frac{1}{R'_{REF} \cdot \left(\frac{C_1 \cdot C_2}{C_1 + C_2} + C_{1,p}\right)} \approx \frac{1}{R'_{REF} \cdot (C_2 + C_{1,p})} \tag{4.59}$$

with $C_{1,p}$ ($\approx 0.2 \cdot C_1$) the parasitic capacitance of the bottom plate of C_1 (cf Table 4.6). Since this pole is much larger than the pole of the OTA it can be assumed in the calculations that the positive OTA input V_+ is grounded (cf Fig. 4.25 (b)).

Analysis shows that the settling-error voltages V_{err,ϕ_i} at the output V_{out} of the amplifier in the two phases are given by

$$V_{err,\phi_2} = \delta_2 \cdot (V_{f,\phi_2} - V_{i,\phi_2}) = \delta_2 \cdot \Delta V_{\phi_2} \tag{4.60}$$

$$V_{err,\phi_1} = \delta_1 \cdot (V_{f,\phi_1} - V_{i,\phi_1}) = \delta_1 \cdot \Delta V_{\phi_1} \tag{4.61}$$

where V_{f,ϕ_i} are the ideal output voltages at the end of the two phases and V_{i,ϕ_i} are the initial output voltages in the two phases. These are determined by the charge redistribution that occurs at the beginning of the phases. Neglecting the switch resistances and parasitic capacitances, and assuming that the OTA is not fast enough to influence the charge redistribution, it can be shown that the initial voltages are equal to

$$V_{i,\phi_2} \approx 0 \tag{4.62}$$

$$V_{i,\phi_1} = \frac{C_{eq}}{C_{eq} + C_{S/H}} V_{in-} + \frac{C_{S/H}}{C_{eq} + C_{S/H}} V_{out,prev} \approx \frac{1}{2} V_{out,prev} \tag{4.63}$$

where

$$C_{eq} = \frac{C_1 \cdot C_2}{C_1 + C_2} \approx C_2 \tag{4.64}$$

and $V_{out,prev}$ is the output voltage during the previous amplification phase ϕ_1. In the calculations ΔV_{ϕ_2} and ΔV_{ϕ_1} are assumed equal to 250 mV and 1.1 V (i.e. the full output-voltage range) respectively, including a safety margin. It can also be shown that these assumptions hold if the parasitic capacitances are included in the model.

δ_2 and δ_1 are equal to

$$\delta_2 = \exp\left(-p_{cl,2} \cdot \frac{\tau_2}{\rho_2}\right) \tag{4.65}$$

$$\delta_1 = \exp\left(-\frac{p_{cl,1}}{1 + \lambda_1} \cdot \frac{\tau_1}{\rho_1}\right) \tag{4.66}$$

where the closed-loop poles $p_{cl,i}$ in the two phases (Fig. 4.25) are given by

$$p_{cl,2} = \frac{g_m}{C_1 + C_2} \tag{4.67}$$

$$p_{cl,1} = \frac{g_m}{C_1} \tag{4.68}$$

and the parameter

$$\lambda_1 = p_{cl,1} \cdot S'_i \cdot C_1 \tag{4.69}$$

models the influence of the input-switch resistance and the strain-gauge resistance on the settling behavior. ρ_2 and ρ_1 are given by

$$\rho_2 = \frac{A \cdot f_{dc2}}{1 + A \cdot f_{dc2}} \tag{4.70}$$

$$\rho_1 = \frac{A \cdot f_{dc1}}{1 + A \cdot f_{dc1}} \tag{4.71}$$

with f_{dc2} and f_{dc1} (cf Eq. (4.53)) the capacitive feedback factors during the reset and amplification phase respectively.

$$f_{dc2} = 1 \tag{4.72}$$

(4.73)

The times τ_2 and τ_1 available for linear settling depend on the slewing behavior of the OTA. Slewing occurs if the following condition is satisfied for the input voltage V_a of the OTA:

$$|V_a| > \frac{I_{SR}}{g_m} = 250 \text{ mV} \tag{4.74}$$

with I_{SR}=44 μA the maximum available output current. τ_2 and τ_1 are given by

$$\tau_2 = t_s - t_{SR,2} = t_s - \frac{\Delta V_{SR,2}}{\frac{I_{SR}}{C_{eq,ol,2}}} \tag{4.75}$$

$$\tau_1 = t_s - t_{SR,1} = t_s - \frac{\Delta V_{SR,2}}{\frac{I_{SR}}{C_{eq,ol,1}}} \tag{4.76}$$

where $t_s \approx \frac{1}{2 \cdot \phi_{sample}}$ represents the total time available during each phase. $t_{SR,i}$ are the times required for potential slewing and ΔV_{SR} the slewing voltage-interval. Note that, because the feedback loop is actually open during the slewing, the equivalent open-loop load capacitances

$$C_{eq,ol,2} = C_1 + C_2 \tag{4.77}$$

$$C_{eq,ol,1} = C_{eq} \tag{4.78}$$

are used for modeling the OTA slewing behavior. Detailed analysis shows that for the implemented amplifier the reduction in available time for linear settling due to slewing is negligible in both phases.

Up till now parasitic capacitances have been omitted. Fig. 4.26 shows the model of the amplifier extended with parasitic capacitances. The capacitances in the C07MA-technology are poly-diffusion (poly–n-well) capacitances. A drawback of these capacitances is the voltage-dependent parasitic capacitance between their bottom plate (i.e. n-well) and the substrate. In

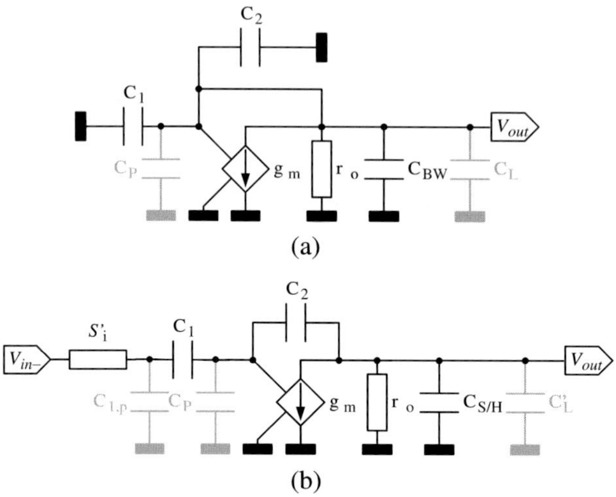

g_m	176 μA/V
r_o	96 MΩ
I_{SR}	44 μA
C_P	500 fF
C_L	100 fF
C_{BW}	250 pF
$C_{S/H}$	12 pF
S'_i	5500 Ω

Figure 4.26: *The resettable gain amplifier including parasitic capacitances (a) during ϕ_2 and (b) during ϕ_1.*

V_{NP}	0 V	0.85 V	1.55 V	3.1 V
$\frac{C_{PAR}}{C_{CAPA}}$	42.7 %	23.7 %	17.8 %	11.8 %

Table 4.6: Overview of the relative parasitic capacitances for different bottom-plate−substrate voltages.

the implemented design bottom-plate parasitic capacitances are avoided at the input nodes of the OTA which are sensitive to potential substrate-noise injection. The parasitic capacitances form an extra load to the OTA (C'_L) or the current sources driving the strain gauges ($C_{1,p}$). The nominal capacitance-per-area C_{CAPA} of the poly-diffusion capacitances is equal to 0.75 fF/μm^2 [Alc 01]. The bottom-plate parasitic capacitance-per-area is given by [San 94]

$$C_{PAR,max} = \frac{C_j}{\left(1 + \frac{V_{NP}}{\phi_j}\right)^{mj}} \tag{4.79}$$

with $C_{j,max}$=3.2e-4 F/m^2, ϕ_j=0.8 V, mj=0.812 [Alc 01] and V_{NP} the voltage between the bottom plate and the substrate. Table 4.6 gives an overview of the relative parasitic capacitances for different bottom-plate−substrate voltages. Because the minimum expected voltage at the output of the amplifier is equal to 1.05 V, a maximum relative error $\frac{C_{PAR,max}}{C_{CAPA}}$ of 25 % is used for the parasitic capacitances in the calculations. The total parasitic capacitance at the output of the amplifier in the amplification phase is equal to

$$C'_L = C_L + 1.25 \cdot C_{S/H} + 0.25 \cdot C_2 \tag{4.80}$$

where C_L is the parasitic output capacitance of the OTA. In Fig. 4.26 C_P represents the parasitic input capacitance of the OTA and $C_{1,p}$ the bottom-plate parasitic capacitance of the input capacitor C_1. Detailed analysis shows that the influence of the parasitic capacitance $C_{1,p}$ is negligible for the implemented amplifier. In the remainder of the calculations this capacitance is omitted in order to reduce the complexity of the final equations and to improve the understanding.

Due to the introduction of the parasitic capacitances and the capacitances C_{BW} and $C_{S/H}$ at the OTA output the equations describing the settling behavior change. The equation for the feedback factor in the amplification phase becomes

$$f_{dc1} = \frac{C_2}{C_1 + C_2 + C_P} \tag{4.81}$$

The closed-loop poles in the two phases are equal to

$$p_{cl,2} = \frac{g_m}{C_{eq,cl,2}} \tag{4.82}$$

$$p_{cl,1} = \frac{g_m}{C_{eq,cl,1}} \tag{4.83}$$

where the equivalent open-loop load capacitances are now given by

$$C_{eq,ol,2} = C_L + C_P + C_1 + C_2 + C_{BW} \tag{4.84}$$

$$C_{eq,ol,1} = (C_1 + C_P)\cdot f_{dc1} + C'_L \tag{4.85}$$

and the equivalent closed-loop load capacitances by

$$C_{eq,cl,2} = \frac{C_{eq,ol,2}}{f_{dc2}} = C_L + C_P + C_1 + C_2 + C_{BW} \tag{4.86}$$

$$C_{eq,cl,1} = \frac{C_{eq,ol,1}}{f_{dc1}} = C_P + C_1 + \frac{C'_L}{f_{dc1}} \tag{4.87}$$

In order to obtain in both phases an output-voltage settling-error smaller than 0.1 % of the wanted voltage accuracy-level $V_{err,\sigma}$, δ_2 and δ_1 must satisfy the following conditions:

$$\delta_2 < 0.1\%\cdot\frac{V_{err,\sigma}}{A_{AMP}\cdot 250\ \text{mV}} \tag{4.88}$$

$$\delta_1 < 0.1\%\cdot\frac{V_{err,\sigma}}{1.1\ \text{V}} \tag{4.89}$$

with (cf Eq. (4.6))

$$V_{err,\sigma} = G\cdot\varepsilon_{err,\sigma}\cdot V_{REF}\cdot A_{AMP} \approx 2.3\ \text{mV} \tag{4.90}$$

Note that the gain of amplifier A_{AMP} is introduced in the condition for δ_2, because the resulting voltage on C_1 at the end of the reset phase is amplified during the next amplification phase,

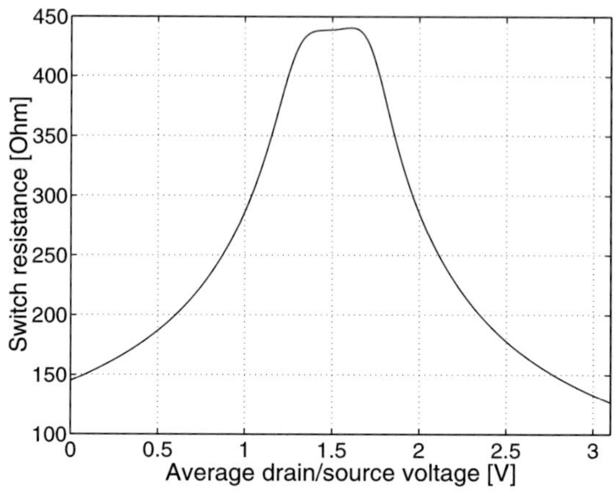

Figure 4.27: *Resistance of the transmission gate sw2,i as a function of the average of the drain and source voltages.*

hence the factor $\frac{1}{A_{AMP}}$ in Eq. (4.88). The times τ_2 and τ_1 for the implemented amplifier are approximately equal to

$$\tau_2 \approx t_s = \frac{1}{2 \cdot \phi_{sample}} \tag{4.91}$$

$$\tau_1 \approx t_s = \frac{1}{2 \cdot \phi_{sample}} \tag{4.92}$$

From these equations, and Eq. (4.65) and Eq. (4.66) follows (with $p_{cl,2}$ and $p'_{cl,1}$ expressed in Hz)

$$\frac{p_{cl,2}}{\phi_{sample}} > \frac{-\ln(\delta_2) \cdot \rho_2}{2\pi \cdot \tau_2 \cdot \phi_{sample}} \approx 5 \tag{4.93}$$

$$\frac{p'_{cl,1}}{\phi_{sample}} = \frac{\frac{p_{cl,1}}{1+p_{cl,1} \cdot S'_i \cdot C_I}}{\phi_{sample}} > \frac{-\ln(\delta_1) \cdot \rho_1}{2\pi \cdot \tau_1 \cdot \phi_{sample}} \approx 4.2 \tag{4.94}$$

For the implemented circuit the ratios $\frac{p_{cl,2}}{\phi_{sample}}$ and $\frac{p'_{cl,1}}{\phi_{sample}}$ are equal to 13.6 and 5 respectively, satisfying Eq. (4.93) and Eq. (4.94).

Detailed analysis also shows that the influence on the settling behavior of the on-resistances of the implemented switches other than the input switches sw2,i is negligible.

4.3.4.5 Switches

All the switches in Fig. 4.20 are transmission gates except for the switches connected to the OTA inputs, clocked by ϕ_2. The reason for this is explained in Section 4.3.4.6. The resistance of a

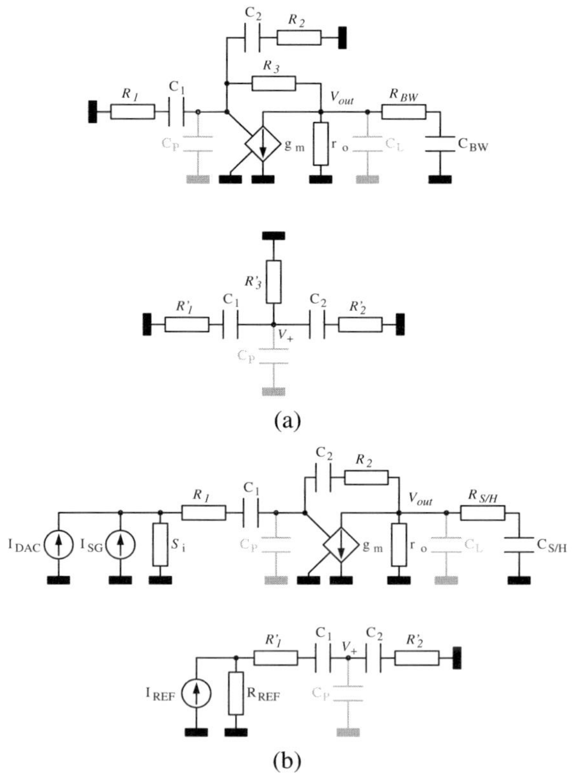

Figure 4.28: *Overview of the switches in the two phases of the amplifier.*

transmission-gate switch, which is composed of a NMOS and a PMOS in parallel, is given by

$$\frac{1}{R_C} = \frac{1}{R_N} + \frac{1}{R_P} \tag{4.95}$$

where the resistances of the NMOS and PMOS can be expressed as [San 94, Mar 99]

$$R_N = \frac{1}{KP_n \cdot \left(\frac{W_n}{L_n}\right) \cdot \left((V_{G,n} - V_{T,n}) - \frac{V_S + V_D}{2}\right)} \tag{4.96}$$

$$R_P = \frac{1}{KP_p \cdot \left(\frac{W_p}{L_p}\right) \cdot \left(\frac{V_S + V_D}{2} - (V_{G,p} - V_{T,p})\right)} \tag{4.97}$$

The switch resistance R_C is thus dependent on the gate voltages $V_{G,i}$, the threshold voltages $V_{T,i}$, and the average of the source and drain voltages V_S and V_D. As an example Fig. 4.27

shows the total resistance of the implemented switch swi,2 (Fig. 4.22) as a function of the average source/drain voltage. The switch is composed of a NMOS with W_n=43.5 μm and L_n=0.7 μm, and a PMOS with W_p=223 μm and L_p=1.2 μm. The maximum resistance of the switch is smaller than 500 Ω.

Fig. 4.28 illustrates the different switches in the two phases of the amplifier. The notations introduced here are also used in the remainder of the text for the noise calculations. The resistances R_1 and R_1' have a maximum value of 500 Ω (cf Fig. 4.27) while the other resistances have a maximum value of 1500 Ω. Note that R_3 and R_3' correspond with the resistances of the single-transistor PMOS switches sw3 and sw3', clocked by the inverse clock $\overline{\phi_2}$, connected to the OTA inputs.

4.3.4.6 Clock feedthrough and charge injection

MOS switches introduce errors due to charge injection, also commonly called clock feedthrough [Joh 97]. These errors are due to unwanted charges being injected into the circuit when the switches (i.e. transistors) turn off. The errors are caused by two mechanisms. The first one is due to the channel charge, which must flow out of the channel region of the transistor to the drain and source junctions. The second one is due to the overlap capacitance between the gate and the source/drain junction.

When a transistor turns off a channel charge equal to [Joh 97]

$$Q_{ch} = W \cdot L \cdot C_{ox} \cdot (V_{GS} - V_T) \tag{4.98}$$

must flow out of the channel region. In [Weg 87] it is explained that the fraction of the channel charge flowing to the source and drain depends on the gate voltage, the threshold voltage, the slope of the gate voltage during switching, the capacitance at the source and drain terminal, and β ($=\frac{W}{L}\mu C_{ox}$). For fast clocks (with a large gate-voltage slope) the charge splits up equally, roughly independent of the capacitance/impedance at the source/drain. For slow clocks the division of the charge over the source and drain depends on the impedance/capacitance seen at this nodes. The largest part of the charge flows to the side with the lowest impedance. From [Weg 87] follows that the clock of the implemented circuit is a fast one, so that it can be assumed that the channel charge splits up approximately equally over the source and the drain.

First the reset phase ϕ_2 is considered. Eq. (4.98) shows that the charge built up in the channel depends on the gate-source voltage as well as the threshold voltage of the transistor. To avoid errors due to the voltage dependence of the injected channel charge, the switches connected to the negative and positive OTA inputs sw3 and sw3' (cf Fig. 4.20) are clocked by the non-delayed clock phase $\overline{\phi_2}$. Because one of the terminals of these switches is connected to ground or virtual ground, the overdrive voltage of these switches is constant from one reset phase to the other. In this way the injected channel charge of these switches is not voltage dependent and can be considered constant from one clock cycle to the other, only giving rise to a small DC offset-voltage.

The single-transistor PMOS switches sw3 and sw3' are clocked by $\overline{\phi_2}$, while the other switches, implemented with transmission gates, are clocked by the delayed clock ϕ_{2d}. The reason for switching off the switches sw3 and sw3' first in the reset phase is to make sure that the charges injected by the other switches have no effect. After switching off sw3 and sw3' one of the terminals of the capacitors C_1 and C_2 is high-impedant, so that no charges can flow to that terminal. The charges injected at the other terminal causes no change in the charge stored on these capacitances, since no low-resistive path exist for charges to flow to the other terminal. Because the charges stored remain the same, the operation of the amplifier is not affected anymore by the extra charges of the switches other than sw3 and sw3'.

Single-transistor PMOS switches are employed instead of transmission gates for the switches sw3 and sw3'. This can be done because one of the terminals of these switches is connected to ground or virtual ground. Although a transmission gate would in principle have no charge injection since the negative charges of the NMOS cancel the positive charges of the PMOS, this is not true in practice. The reason for this is that the NMOS and PMOS transistors and the clock signals controlling them are not perfectly matched. In [Pet 86] the importance of the skew time between the PMOS and NMOS clock waveforms has been investigated. It is shown that although complementary switches are reported in literature to have inherent channel charge cancellation abilities, this cancellation is very sensitive to clock skew. Practical switched-capacitor systems, having complementary switches, may show higher amounts of clock feedthrough than identical systems having single-transistor switches. To avoid these problems single transistor (PMOS) switches, driven by the same clock, have been used for the switches connected to the OTA inputs, which are critical for charge injection.

As already mentioned above, another error source is present in MOS switches [Gee 01, Joh 97]. Due to the overlap capacitance between the gate and the source/drain region, the voltage at the latter nodes changes when the gate signal switches. Under the assumption that the clock switches infinitely fast, all the charge from the clock feedthrough ends up at the source/drain terminal instead of flowing through the switch while the transistor is not yet off. The resulting voltage step for a high-impedance node can be expressed as

$$\Delta V = \frac{C_{ov}}{C_{ov} + C_i} V_{CLK} \tag{4.99}$$

where C_{ov} denotes the overlap capacitance, C_i the capacitance at the source/drain terminal, and V_{CLK} the amplitude of the clock signal. For the implemented circuit the clock feedthrough due to the overlap capacitances of the switches gives rise to small glitches at the end of ϕ_2 until the source/drain terminals of the switches are charged/discharged during the next amplification phase ϕ_1 by low-impedance nodes. This is true for all switches except for sw3 and sw3', which cause an error in the output voltage at the end of the next amplification phase ϕ_1. When they are shut off the overlap capacitors between their gates and C_1 and C_2 pull some charge out of C_1 and C_2 causing offset errors.

In [Mar 82] is explained that these errors can be minimized by making the capacitances seen at the negative and positive OTA input the same and by making the switches sw3 and sw3' the same. In this way the clock feedthrough at both sides gives rise to the same error voltage (cf Eq. (4.99)). This is especially true for fast clocks where the channel charge splits up equally, roughly independent of the impedances seen at the source and drain side of the switches. The principle introduced in [Mar 82] has been applied in the implemented circuit. At the end of the reset phase the single-transistor PMOS switches clocked by $\overline{\phi}_2$ both have one terminal connected to one of the OTA inputs with the same total capacitance equal to C_1+C_2. When ϕ_2 goes low $\overline{\phi_2}$ goes high switching off the PMOS switches. The injected charges ideally cause an error voltage, equal at both sides of the OTA, so that the resulting output-voltage error only depends on the CMRR of the OTA. In practice however the resulting error also depends on the matching of the switches and the capacitors. Because for the implemented amplifier C_i in Eq. (4.99) is equal to C_1+C_2, which is much larger than C_{ov}, the resulting error due to mismatch is negligible (cf Eq. (4.99)).

Up till now only the reset phase of the amplifier has been considered. Also in the amplification phase errors are introduced due to charge injection and clock feedthrough of the switches. Moreover, the channel-charge injection of the input switches sw2,i in the amplification phase is dependent on the input voltage V_{in-}. These errors however don't give rise to an error in the output voltage V_{out}, since the switch connected to the sample-and-hold capacitance $C_{S/H}$ and virtual ground of the S/H circuit switches off first (cf Fig. 4.31). Like in the case of the amplifier, only the charges injected by the switches connected to the inputs of the S/H OTA are important at the end of ϕ_1. In this way the voltage-dependent charge injection due to the input switches swi2,i does not play a role, so that no gain and distortion errors [Joh 97] are introduced by theses switches.

4.3.4.7 Noise

This subsection presents the calculation of the equivalent input-referred noise resulting from the various noise components of the amplifier. An overview of all the noise sources in the two phases is shown in Fig. 4.29. The noise appearing at the output of the switched-capacitor amplifier during ϕ_1 is due to two different propagation methods: direct broadband noise and sample-and-hold noise [Gob 83, Gee 01]. The direct broadband noise is due to noise sources with direct coupling to the output during ϕ_1. The sampled noise component is due to the sampling of the direct broadband noise on the capacitances C_1 and C_2 (at both sides of the OTA) at the end of the previous reset phase ϕ_2. Since the bandwidth of the broadband noise is much larger than the sampling frequency, noise aliasing takes place during the sampling operation.

Three different types of noise sources can be distinguished in Fig. 4.29: the noise components related to the current sources (in ϕ_1), the noise sources related to the switches, and the noise contribution of the OTA in the two phases. The noise components related to the current sources are due to the transistors of the current mirrors used to derive the currents I_{REF}, I_{SG} and I_{DAC} from I_{SOURCE}. Also a common-mode noise source $\overline{di^2}_{I_{SOURCE}}$ related to the reference current source is included in Fig. 4.29. Detailed analysis shows that the contribution of this noise source as well as the 1/f noise contributions of the OTA and the other components can be neglected. Therefore only the white noise components are considered in the remainder of this section. The power spectral density of the white current-noise of a switch with a resistance R is given by

$$\overline{di^2}_R = \frac{4kT}{R} \tag{4.100}$$

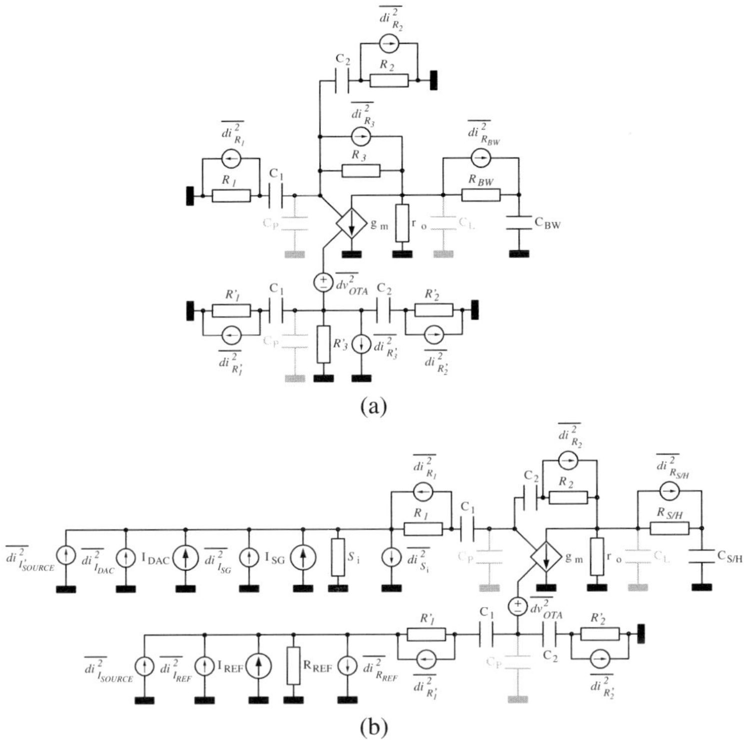

Figure 4.29: *Overview of the different noise components (a) during ϕ_2 and (b) during ϕ_1.*

and the power spectral density of the white voltage-noise of the OTA by

$$\overline{dv^2}_{OTA} = \frac{8kT}{3g_{m1}}\gamma \tag{4.101}$$

where γ is the noise excess factor of the OTA [San 94], which is defined as the ratio of the equivalent input noise of the OTA to the noise of the input transistor. For the implemented OTA (cf Fig. 4.30) γ is equal to

$$\gamma = 2(1 + \frac{g_{m7}}{g_{m1}} + \frac{g_{m9}}{g_{m1}}) \approx 7.5 \tag{4.102}$$

In order to calculate the total input-referred noise at the end of ϕ_1 the following procedure is followed. First the *sampled noise contributions* of the noise components at the end of ϕ_2 are calculated. For every noise source the resulting noise voltages, sampled on the capacitances C_1 and C_2 at both sides of the OTA, are determined. Note that the noise contributions on different capacitances due to the same noise source are correlated.

To refer the noise voltages to the input the noise voltages sampled on C_2 at the end of ϕ_2 must be divided by $\frac{C_1}{C_2}$. This follows from the law of charge conservation. The charges stored

T(s)	BW_{eq}
$\frac{1}{s+p_1}$	$\frac{1}{4} \cdot p_1$
$\frac{1}{(s+p_1)\cdot(s+p_2)}$	$\frac{1}{4} \cdot \frac{p_1 \cdot p_2}{p_1+p_2}$
$\frac{s+z}{(s+p_1)\cdot(s+p_2)}$	$\frac{1}{4} \cdot \frac{(p_1 \cdot p_2+z^2)\cdot p_1 \cdot p_2}{(p_1+p_2)\cdot z^2}$
$\frac{(s+z_1)}{(s+p_1)\cdot(s+p_2)\cdot(s+p_3)}$	$\frac{1}{4} \cdot \frac{[p_1 \cdot p_2 \cdot p_3+(p_1+p_2+p_3)\cdot z_1^2]\cdot p_1 \cdot p_2 \cdot p_3}{(p_3+p_2)\cdot(p_2+p_1)\cdot(p_3+p_1)\cdot z_1^2}$
$\frac{(s+z_1)\cdot(s+z_2)}{(s+p_1)\cdot(s+p_2)\cdot(s+p_3)}$	$\frac{[(p_1 \cdot p_2 \cdot p_3)\cdot(p_1 \cdot p_2+p_1 \cdot p_3+p_2 \cdot p_3)+(p_3 \cdot p_1 \cdot p_2)\cdot(z_1^2+z_2^2)+(p_1+p_2+p_3)\cdot z_1^2 \cdot z_2^2]\cdot p_1 \cdot p_2 \cdot p_3}{4 \cdot z_2^2 \cdot z_1^2 \cdot (p_3+p_2)\cdot(p_1+p_3)\cdot(p_2+p_1)}$
$\frac{s}{(s+p_1)\cdot(s+p_2)}$	$\frac{1}{4} \cdot \frac{1}{p_1+p_2}$
$\frac{s^2}{(s+p_1)\cdot(s+p_2)\cdot(s+p_3)}$	$\frac{1}{4} \cdot \frac{p_1 \cdot p_2+p_3 \cdot p_1+p_3 \cdot p_2}{(p_3+p_1)\cdot(p_3+p_2)\cdot(p_2+p_1)}$
$\frac{s\cdot(s+z_1)}{(s+p_1)\cdot(s+p_2)\cdot(s+p_3)}$	$\frac{1}{4} \cdot \frac{p_2 \cdot p_3+p_1 \cdot p_3+p_2 \cdot p_1+z_1^2}{(p_3+p_2)\cdot(p_3+p_1)\cdot(p_1+p_2)}$

Table 4.7: Overview of the equivalent bandwidths for different tranfert functions.

on C_2 at the end of ϕ_2 remain on C_2 during the next amplification phase, while the charges on C_1 are transferred to C_2. In this way the noise voltage sampled on C_1 is amplified by a factor $\frac{C_1}{C_2}$ to the output of the amplifier while the noise voltage sampled on C_2 is not amplified. Detailed analysis shows that due to this the total noise contribution due to the noise voltages sampled on C_2 is small in comparison with the one due to the noise voltages sampled on C_1.

Next the *direct noise contributions* of the noise components to the output of the amplifier in ϕ_1 are calculated. To refer the output noise voltages to the input they have to be divided by the amplifier gain $\frac{C_1}{C_2}$.

Because of the 'virtual ground' at the negative OTA input, the noise sources at the positive OTA input side give rise to noise contributions at the positive side as well as at the negative side. Ideally, if the bandwidth limitation of these noise components at the two sides of the OTA would be the same, the resulting sampled noise voltages would be the same and they would cancel each other. Because of the different noise bandwidths however, the sampled noise voltages at the end of ϕ_2 are assumed uncorrelated in the calculations, giving rise to a slightly overestimated input-referred noise.

Small-signal analysis is applied to determine the transfert functions T(s) from the different noise voltage/current-sources to the voltages across the capacitances (in ϕ_2) and to the amplifier output (in ϕ_1). The rms value of the resulting noise voltage for a noise source $\overline{dv^2}$ or $\overline{di^2}$ is given by $\sqrt{N_{tot,i}}$ with

$$N_{tot,i} = \int_0^\infty \overline{dv,i^2} \cdot |T(s)|^2_{s=j\cdot 2\pi \cdot f} df = \overline{dv,i^2} \int_0^\infty |T(s)|^2_{s=j\cdot 2\pi \cdot f} df = \overline{dv,i^2} \cdot BW_{eq} \quad [V^2] \qquad (4.103)$$

Noise source	ϕ_2	ϕ_1
R_1	1.69e-12	/
R_2	7.96e-14	2.09e-14
R_3	1.03e-12	/
R'_1	4.64e-14 / 1.43e-12	/
R'_2	2.30e-15 / 6.40e-14	6.40e-15
R'_3	8.82e-13 / 4.17e-12	/
S'_i	/	1.80e-12
R'_{REF}	/	3.15e-12
R_{BW}	1.32e-12	/
$R_{S/H}$	/	1.81e-14
$\overline{dv^2}_{OP}$	2.55e-11	9.56e-12
$\overline{di^2}_{ISG}$	/	6.70e-12
$\overline{di^2}_{IDAC}$	/	5.98e-12
$\overline{di^2}_{IREF}$	/	1.31e-11
Total	**3.62e-11**	**4.03e-11**

Table 4.8: Overview of the input-referred total-integrated-noise-powers $N'_{tot,i}$ for the various noise sources in the two phases (T=314 K).

the total integrated noise power. Note that in Eq. (4.103) $\overline{dv,i^2}$ is assumed constant which is the case for white noise sources. Table 4.7 gives an overview of the equivalent bandwidths of some important transfer functions appearing in the calculations. Note that the table gives the normalized equivalent bandwidth, where T(s) is replaced by $\frac{T(s)}{T(0)}$ in Eq. (4.103), except for the transfer functions with at least one zero equal to 0.

The total input-referred rms noise voltage $V_{rms,in}$ is found with [San 94]

$$V_{rms,in} = \sqrt{\sum_{i=1}^{N} N'_{tot,i}} \tag{4.104}$$

where $N'_{tot,i}$ are the individual input-referred total-integrated-noise-powers. An overview of the latter for the implemented circuit is given in Table 4.8. The noise contributions at the end of ϕ_2 due to R'_i at the negative OTA input side are listed before the noise contributions at the positive side. The total input-referred rms noise voltage is equal to 8.75 μV, which is $0.27 \cdot V_{err,\sigma,in}$ with $V_{err,\sigma,in} (= 32.9\mu V)$ the input-referred wanted voltage accuracy-level (cf Eq. (4.90)).

4.3.4.8 Distortion

In this subsection two sources of distortion are investigated: the voltage dependence of the capacitances and the output-voltage dependent OTA gain A. Similar calculation methods as in [Lee 85, Gee 01, Mar 99] are used to analyze the distortion components.

The first source of distortion is the voltage dependence of the capacitances. The value of the capacitors varies with the voltage applied to their terminals. For the poly-diffusion capacitances of the C07MA technology the voltage dependence is given by

$$C_i(v) = C_{i0} \cdot (1 + a_1 \cdot v) \tag{4.105}$$

where C_{i0} is the nominal capacitance value when the capacitor carries no charge and a_1 is the linear coefficient of the capacitors with a maximum value of 50 ppm/V which value is used in the calculations. The voltage applied over the terminals of the capacitor is indicated by v.

The charge conservation in ϕ_1 for an ideal amplifier (with $V_{off} = 0$ V) can be written as (cf Eq. (4.44)-Eq. (4.47))

$$(V_{in+} - V_{in-}) \cdot C_1 - V_{out} \cdot C_2 = V_{in} \cdot C_1(V_{in}) - V_{out} \cdot C_2(V_{out}) = 0 \tag{4.106}$$

with V_{in} the difference between the input voltages. Since a_1 is much smaller than 1, this equation can also be written as

$$V_{out} = \frac{C_1}{C_2} \cdot V_{in} \cdot (1 + a_1 \cdot V_{in}) \cdot (1 - a_1 \cdot V_{out}) \tag{4.107}$$

In order to calculate the distortion a sinusoidal input signal is assumed with an amplitude V_i and an angular frequency ω_i.

$$V_{in} = V_i \cdot \cos(\omega_i \cdot t) \tag{4.108}$$

The output signal of the amplifier contains the amplified input signal and small distortion components. In first order these components can be neglected and the output can be represented by

$$V_{out} = V_o \cdot \cos(\omega_i \cdot t) \tag{4.109}$$

with an amplitude V_o. By substitution of Eq. (4.108) and Eq. (4.109) into the right-hand side of Eq. (4.107) the latter can be expanded into a Fourier series.

$$V_{out} = HD_3 \cdot \cos(3 \cdot \omega_i \cdot t) + HD_2 \cdot \cos(2 \cdot \omega_i \cdot t) + S \cdot \cos(\omega_i \cdot t) + DC \tag{4.110}$$

with

$$HD_3 = -\frac{1}{4} \cdot \frac{C_1}{C_2} \cdot V_i^2 \cdot a_1^2 \cdot V_o \tag{4.111}$$

$$HD_2 = -\frac{1}{2} \cdot \frac{C_1}{C_2} \cdot V_i \cdot a_1 \cdot Vo + \frac{1}{2} \cdot \frac{C_1}{C_2} \cdot V_i^2 \cdot a_1 \tag{4.112}$$

$$S = \frac{C_1}{C_2} \cdot V_i - \frac{3}{4} \cdot \frac{C_1}{C_2} \cdot V_i^2 \cdot a_1^2 \cdot V_o \approx \frac{C_1}{C_2} \cdot V_i \tag{4.113}$$

$$DC = -\frac{1}{2} \cdot \frac{C_1}{C_2} \cdot V_i \cdot a_1 \cdot V_o + \frac{1}{2} \cdot \frac{C_1}{C_2} \cdot V_i^2 \cdot a_1 \tag{4.114}$$

where HD_3 and HD_2 are the third and the second harmonic distortion terms, S is the fundamental term and DC the DC-term. Note that for the amplifier V_o is given by

$$V_o = \frac{C_1}{C_2} \cdot V_i \tag{4.115}$$

so that the ratio of the signal S to the third and second order harmonic terms can be expressed as

$$SHD_3 = -20 \cdot \log_{10}\left(\frac{1}{4}\cdot\frac{C_1}{C_2}\cdot V_i^2 \cdot a_1^2\right) \tag{4.116}$$

$$SHD_2 = -20 \cdot \log_{10}\left(\frac{1}{2}\cdot\frac{C_1 - C_2}{C_2}\cdot |V_i| \cdot a_1\right) \approx -20 \cdot \log_{10}\left(\frac{1}{2}\cdot\frac{C_1}{C_2}\cdot |V_i| \cdot a_1\right) \tag{4.117}$$

This shows that SHD_2 and SHD_3 are proportional to respectively the amplitude of the input/output signal and the square of the amplitude of the input/output signal. For the maximum input signal, corresponding with $\varepsilon_{max,min}$, the HD_3-term for the implemented circuit corresponds with a strain equal to 3.23e-16 strain and the HD_2-term with a strain of 6.55e-9 strain, showing that both distortion terms are negligible. Also the DC-term in Eq. (4.110) is negligible.

The second source of distortion is the output-voltage dependent OTA gain A. The gain of the OTA is not fixed but depends on the input and output voltages of the OTA. The dependency on the input voltage can be neglected for the switched-capacitor amplifier, since the input voltage settles to the same constant voltage V'_{in+} (cf Eq. (4.48)) in each amplification phase ϕ_1. However, the output voltage of the OTA in ϕ_1 is equal to the input-signal difference V_{in} amplified by $\frac{C_1}{C_2}$. Therefore, it varies significantly from cycle to cycle and influences the output resistance of the OTA and thus the gain A of the OTA, which results in distortion.

When the output voltage increases, the drain-source voltage V_{DS} of the output transistors decreases so that they come closer to the linear operation region resulting in a reduction of their output impedance and and thus of the gain of the OTA. The non-linear gain of the OTA can be modeled by a Taylor series of second order.

$$A(v) = A_0 \cdot (1 + a_1 \cdot v + a_2 \cdot v^2) \tag{4.118}$$

where v is the output voltage of the OTA. Note that a_2 has a negative value since the gain decreases as the output swing increases.

In Fig. 4.24 (b) the model for the switched-capacitor amplifier in ϕ_1 with a finite OTA gain is shown. By neglecting the parasitic capacitances and assuming $V_{off} = 0$ V the charge conservation can be expressed as

$$V_{out} = (V_{in+} - V_{in-}) \cdot \frac{C_1}{C_2} + V_a \cdot (1 + \frac{C_1}{C_2}) = V_{in} \cdot \frac{C_1}{C_2} + V_a \cdot (1 + \frac{C_1}{C_2}) \tag{4.119}$$

with V_{in} the difference between the input voltages and V_a the input voltage of the OTA given by

$$V_a = -\frac{V_{out}}{A_0} \cdot (1 - a_1 \cdot V_{out} - a_2 \cdot V_{out}^2) \tag{4.120}$$

Using similar calculations as above the following signal-to-harmonic-distortion ratios can be

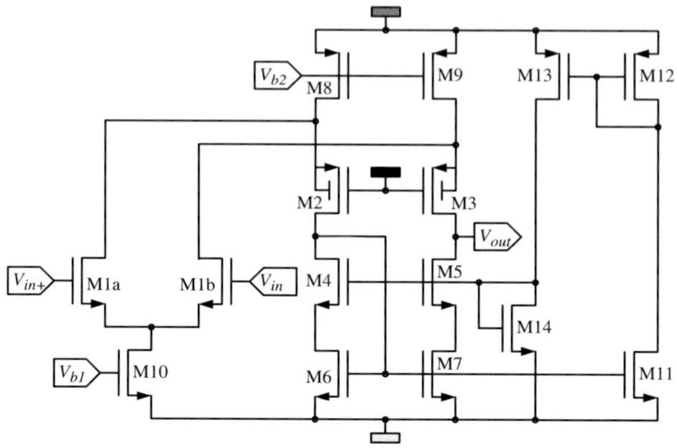

Figure 4.30: *Implemented wide-swing folded-cascode OTA.* V_{b1}*=1.120 V and* V_{b2}*=2.153 V (cf Table A.4).*

calculated

$$\mathrm{SHD_3} = -20\cdot\log_{10}\left(\frac{1}{4}\cdot\frac{|a_2|\cdot V_i^2\cdot\left(1+\frac{C_1}{C_2}\right)\cdot\left(\frac{C_1}{C_2}\right)^2}{A_0}\right) \tag{4.121}$$

$$\mathrm{SHD_2} = -20\cdot\log_{10}\left(\frac{1}{2}\cdot\frac{|a_1|\cdot V_i\cdot\left(1+\frac{C_1}{C_2}\right)\cdot\frac{C_1}{C_2}}{A_0}\right) \tag{4.122}$$

These equations show that the distortion components can be suppressed by increasing the gain of the OTA.

For the implemented wide-swing folded-cascode OTA , shown in Fig. 4.30, a_1=-8.13e-3, a_2=-2.13e-1 and $A_0 \approx 16926$. For the maximum input signal, corresponding with $\varepsilon_{max,min}$, the HD_3-term now corresponds with a strain equal to 8.1e-9 strain and the HD_2-term with a strain of 4.53e-9 strain, showing that also these distortion terms are negligible for the implemented circuit.

4.3.4.9 CMRR and PSRR

In this section the (DC) Common Mode Rejection Ratio (CMRR) and Power Supply Rejection Ratio (PSRR) of the implemented wide-swing folded-cascode OTA are investigated. Due to the cascodes the systematic $CMRR_s$ ($\approx$ 127.3 dB) is negligible in comparison with the random $CMRR_r$, which is approximately equal to

$$\mathrm{CMRR_r} \approx \frac{g_{m1}}{\frac{\sigma_{g_{m1}}}{g_{m1}}\cdot\frac{g_{o10}}{2}} \approx 95.5\ \mathrm{dB} \tag{4.123}$$

	DC	2 kHz
CMRR	127.3 dB	117.3 dB
$PSRR_{ss}$	85.4 dB	85.4 dB
$PSRR_{dd}$	88.9 dB	88.9 dB

Table 4.9: Overview of the simulated CMRR and $PSRR_{ss,dd}$ without mismatch at DC and at ϕ_{sample} (2 kHz).

Note that the mismatch in the bulk transconductances g_{mb1} has been negliged, because of its small value (0.06 %) in comparison with $\frac{\sigma_{g_{m1}}}{g_{m1}}$, which is given by

$$\frac{\sigma_{g_{m1}}}{g_{m1}} \approx \frac{\sigma_{offset}}{V_{GS1} - V_{T1}} \approx 0.4\ \% \tag{4.124}$$

where σ_{offset} is the equivalent input-referred offset (cf Eq. (4.54)).

The $PSRR_{ss}$ is approximately equal to

$$PSRR_{ss} \approx \frac{g_{m1}}{\frac{g_{o2} \cdot g_{o8}}{g_{m2}} + \frac{g_{o5} \cdot g_{o7}}{g_{m5}} + \frac{\sigma_{g_{m1}}}{g_{m1}} \cdot \frac{g_{o10}}{2}} \approx 83.5\ \text{dB} \tag{4.125}$$

If the mismatch term $\frac{\sigma_{g_{m1}}}{g_{m1}} \cdot \frac{g_{o10}}{2}$ is not included, a $PSRR_{ss}$ of 86 dB is found. This corresponds well with the simulated value (without mismatch), i.e. 85.4 dB. The $PSRR_{dd}$ on the other hand can be expressed as

$$PSRR_{dd} \approx \frac{g_{m1}}{g_{o9} - g_{o8} \cdot \frac{g_{m7}}{g_{m6}} + \sigma_{g_{o8}} + g_{o8} \cdot \frac{\sigma_{g_{m6}}}{g_{m6}}} = 87.3\ \text{dB} \tag{4.126}$$

where (cf Eq. (4.55))

$$\sigma_{g_{o8}} \approx g_{o8} \cdot \frac{\sigma_{I_{DS8}}}{I_{DS8}} = g_{o8} \cdot \left(\frac{A^2_{\beta,p}}{W_8 \cdot L_8} + \frac{1}{W_8 \cdot L_8} \cdot \frac{4 \cdot A^2_{VT,p}}{(V_{GS8} - V_{T8})^2} \right)^{\frac{1}{2}} = 0.8\ \text{nS} \tag{4.127}$$

$$\frac{\sigma_{g_{m6}}}{g_{m6}} \approx \frac{\sigma_{M6}}{V_{GS6} - V_{T6}} = 0.09\ \% \tag{4.128}$$

If the mismatch terms are excluded a $PSRR_{dd}$ of 88.8 dB is found. Table 4.9 gives an overview of the simulated CMRR and $PSRR_{ss,dd}$ without mismatch at DC and at ϕ_{sample}, showing that the calculated values (at DC) correspond well with the simulated ones.

The closed-loop transfer functions $CMR_{cl,i}$ and $PSR_{cl,i}$ in ϕ_i for a common-mode input signal v_{cm} and a power-supply signal $v_{ss,dd}$ to the output V_{out} of the resettable gain amplifier are respectively given by [San 94]

$$CMR_{cl,i} = \frac{V_{out}}{v_{cm}} = \frac{1}{CMRR \cdot f_{dci}} \tag{4.129}$$

$$PSR_{cl,i} = \frac{V_{out}}{v_{ss,dd}} = \frac{1}{PSRR_{ss,dd} \cdot f_{dci}} \tag{4.130}$$

	ϕ_2	ϕ_1
CMR_{cl}	95.5 dB	58.4 dB
$PSR_{ss,cl}$	83.5 dB	48.4 dB
$PSR_{dd,cl}$	87.3 dB	51.9 dB

Table 4.10: Closed-loop transfer function due to CMRR and $PSRR_{ss,dd}$ in the two phases.

Error	ϕ_2	ϕ_1
CMRR	23.6 μV	1.57 μV
$PSRR_{ss}$	94.9 μV	94.9 μV
$PSRR_{dd}$	61.3 μV	61.3 μV

Table 4.11: Maximum errors due to the DC CMRR and $PSRR_{ss,dd}$ in the two phases.

with f_{dci} the feedback factor in ϕ_i. Table 4.10 shows an overview of the closed-loop transfer functions due to CMRR and $PSRR_{ss,dd}$ in the two phases including mismatch.

The employed batteries are silver oxide batteries [Ene], which provide a stable operating voltage until the end of discharge. The maximum variation from 1.55 V of the operating voltage is limited to $\pm$20 mV. This means that in ϕ_1 the maximum common-mode input-signal change v_{cm} equals (cf Eq. (4.14))

$$v_{cm} = \frac{\Delta I_{SOURCE}}{\Delta v_{DD,SS}} \cdot 9310\ \Omega = 1.31\ \text{mV} \tag{4.131}$$

with $\Delta v_{DD,SS} = \pm 40$ mV, resulting in an output voltage V_{out} of 1.57 μV. In the reset phase ϕ_2 on the other hand the maximum common-mode input signal varies with $\pm$20 mV. This results in an error equal to 0.34 μV which must be amplified by $\frac{C_1}{C_2}$, because an error voltage on C_1 in ϕ_2 is amplified by this factor during the next amplification phase ϕ_1.

The results for the maximum PSRR-related errors in the two phases are given in Table 4.11. For the calculation of the PSRR-related errors $\Delta v_{DD,SS}$ has been assumed $\pm$20 mV. So far the effect of the CDS technique employed in the resettable gain amplifier has not been considered. Besides cancelling the input offset voltage, the CDS technique also improves the PSRR. The PSRR of an amplifier is a measure of how much the amplifier's output voltage changes with changes in the power supply. This is essentially a change in input offset voltage with the power supply voltage, so that also the effect of low-frequent power supply variations is greatly reduced by the CDS technique. This results in small PSRR-related errors compared with the ones given in Table 4.11, where this effect has not been taken into account [Mur 00].

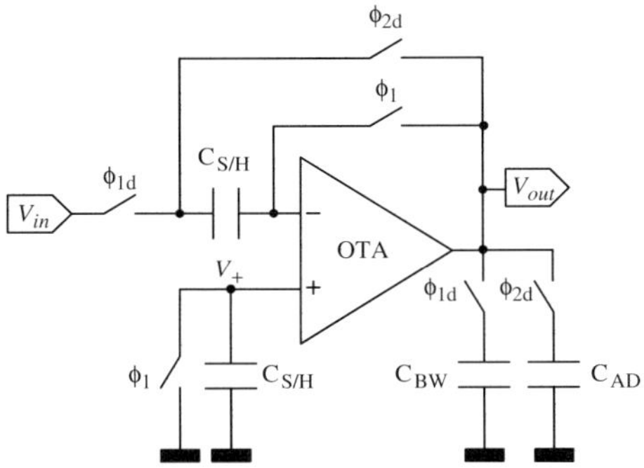

Figure 4.31: *Switched-capacitor sample-and-hold with offset-cancellation.*

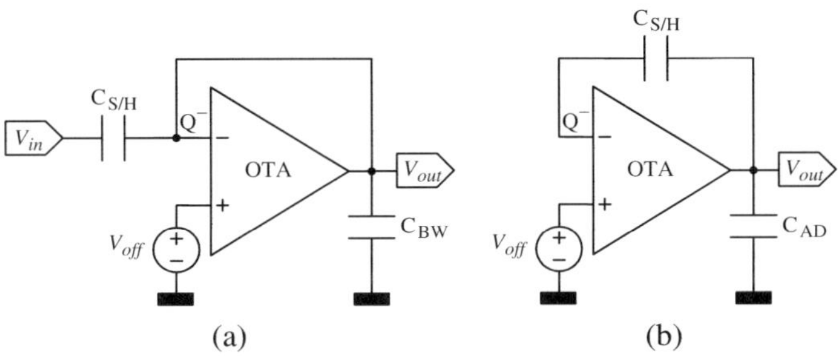

Figure 4.32: *The sample-and-hold circuit (a) during the sample phase ϕ_1 and (b) during the hold phase ϕ_2.*

4.3.5 S/H

4.3.5.1 Operating principle

The next building block is the sample-and-hold S/H. The S/H circuit, depicted in Fig. 4.31, also exploits the CDS technique to cancel the OTA's offset voltage V_{off}. To understand the operation of the S/H, the equivalent circuits in the sample phase ϕ_1 and the hold phase ϕ_2 are shown in Fig. 4.32, where V_{MM} is assumed grounded (cf Section 4.3.4.1). Note that the voltage at the positive input terminal (V_+) of the OTA is 0 V in the two phases. This follows directly from applying the law of charge conservation at this node. It would also be possible to connect the positive OTA input terminal directly to V_{MM}, but by adding an extra switch and a capacitance at the positive OTA input the effects of clock feedthrough are reduced, as explained in Section 4.3.4.6. Also non-delayed clocks $\phi 1$ are employed for the switches connected to the OTA inputs to make sure that the clock feedthrough of the other switches has no effect and the resulting clock-feedthrough error is constant from cycle to cycle, independent of the input voltage. By applying the law of

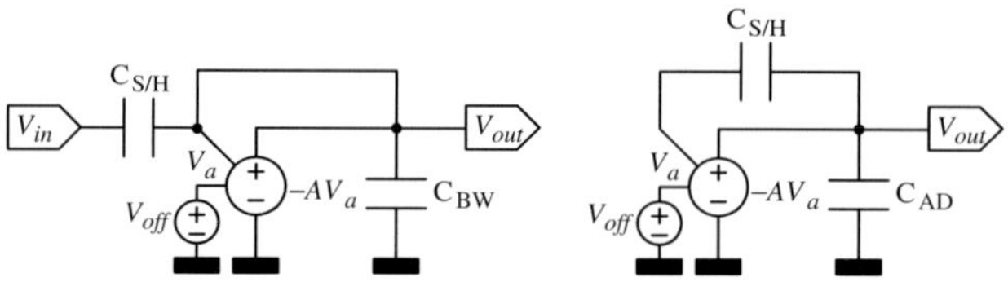

Figure 4.33: *The sample-and-hold circuit considering an OTA with a finite gain A (a) during the sample phase ϕ_1 and (b) during the hold phase ϕ_2.*

charge conservation at the negative OTA input

$$Q^-_{\phi_1} = Q^-_{\phi_2} \Rightarrow (V_{off} - V_{in,\phi_1}) = (V_{off} - V_{out,\phi_2}) \tag{4.132}$$

the input-output relation can be found which is equal to

$$V_{out,\phi_2} = V_{in,\phi_1} \tag{4.133}$$

From Eq. (4.133) follows that the OTA's offset V_{off} is cancelled by the S/H circuit.

4.3.5.2 Finite OTA gain

In this section the influence of the OTA's finite gain on the performance of the S/H is investigated. The two phases of the S/H are reconsidered in Fig. 4.33 with an OTA having a finite gain -A. The equations for $Q^-_{\phi_1}$ and $Q^-_{\phi_2}$ become

$$Q^-_{\phi_1} = (V_{off} \cdot \frac{A}{1+A} - V_{in,\phi_1}) \cdot C_{S/H} \tag{4.134}$$

$$Q^-_{\phi_2} = (V_{off} - \frac{V_{out,\phi_2}}{A} - V_{out,\phi_2}) \cdot C_{S/H} \tag{4.135}$$

By applying the law of charge conservation the following input-output relation

$$V_{out,\phi_2} = \frac{V_{in,\phi_1}}{1+\frac{1}{A}} + \frac{V_{off}}{A} \tag{4.136}$$

is found. The gain of the implemented OTA, which uses a folded-cascode topology (cf Fig. 4.30) with PMOS input transistors, equals approximately 22000. This results in a static error ε_s equal to 0.0045 %, corresponding with a maximum error of ± 0.09 μstrain for the maximum/minimum strain $\varepsilon_{max,min}$. Like in the case of the amplifier this does not impose a problem, because this gain error is calibrated for.

From Eq. (4.136) follows that the resulting offset error in the output is reduced by a factor equal to the gain of the OTA. The input-referred offset σ_{offset} of the implemented OTA, calculated with Eq. (4.54), equals 1.8 mV. This offset voltage is reduced by A, so that the final error in the output due to the offset voltage equals approximately 82 nV, which is negligible.

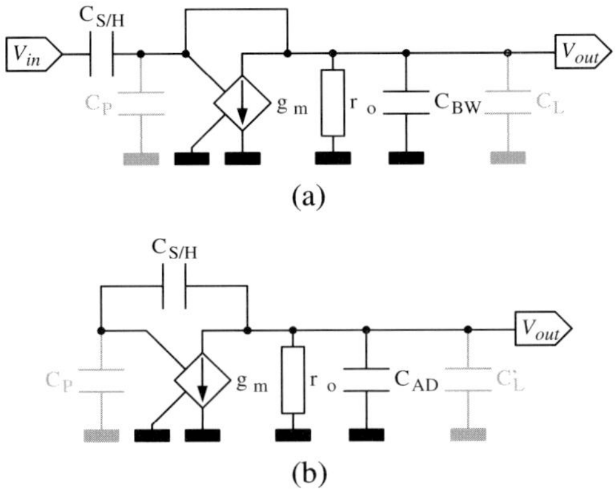

g_m	46.9 μA/V
r_o	469 MΩ
I_{SR}	11.732 μA
C_P	270 fF
C_L	15 fF
C_{BW}	250 pF
$C_{S/H}$	12 pF
C_{AD}	153.6 pF

Figure 4.34: *The sample-and-hold including parasitic capacitances (a) during ϕ_1 and (b) during ϕ_2.*

4.3.5.3 Settling behavior

In this section the settling behavior of the S/H is investigated. Therefore the two phases of the S/H are reconsidered in Fig. 4.34 with the OTA modeled by a transconductance g_m and an output resistance r_o. Detailed analysis shows that the output-voltage errors in the two phases are given by Eq. (4.60) and Eq. (4.61) where the initial output voltages V_{i,ϕ_i}, determined by charge conservation, are approximately equal to

$$V_{i,\phi_1} \approx \frac{C_{S/H} \cdot (V_{in} - V_{out,prev})}{C_{S/H} + C_{BW}} < \frac{1}{20} \cdot 1.1 \text{ V} \tag{4.137}$$

$$V_{i,\phi_2} \approx V_{out,prev} \tag{4.138}$$

The final voltages V_{f,ϕ_1} and V_{f,ϕ_2} are equal to respectively 0 V and V_{in,ϕ_1}. This means that the maximum ΔV_{ϕ_1} and ΔV_{ϕ_2} are approximately equal to 250 mV and 1.1 V. Detailed analysis shows that these assumptions hold if the parasitic capacitances are included.

The closed-loop poles $p_{cl,i}$ in the two phases are given by Eq. (4.82), Eq. (4.83), Eq. (4.86) and Eq. (4.87) where the equivalent open-loop capacitances are now equal to

$$C_{eq,ol,1} = C_L + C_P + C_{S/H} + C_{BW} \tag{4.139}$$

$$C_{eq,ol,2} = C_P \cdot f_{dc2} + C'_L + C_{AD} \tag{4.140}$$

and the feedback factors are given by

$$f_{dc1} = 1 \tag{4.141}$$

$$f_{dc2} = \frac{C_{S/H}}{C_{S/H} + C_P} \tag{4.142}$$

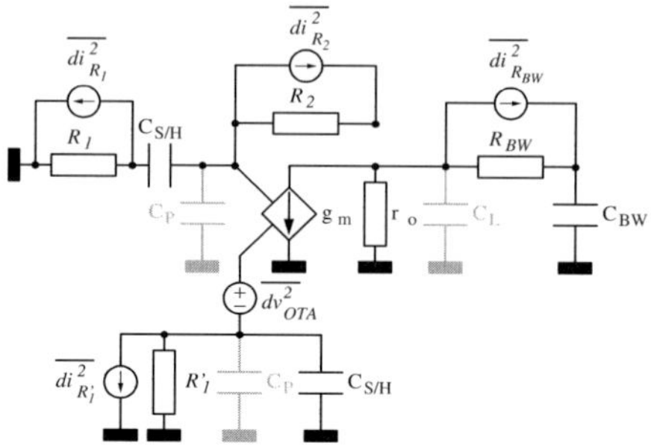

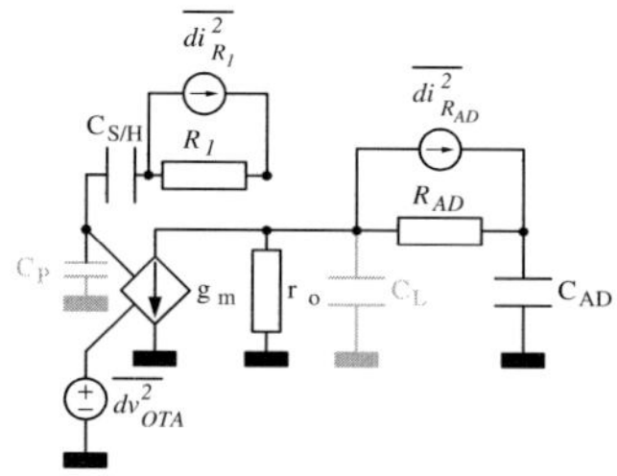

Figure 4.35: *The different noise components (a) during ϕ_1 and (b) during ϕ_2.*

The maximum total parasitic capacitance C'_L at the output of the amplifier in the sample phase ϕ_1 equals

$$C'_L = C_L + 0.25{\cdot}C_{S/H} + 0.25{\cdot}C_{AD} \tag{4.143}$$

As explained in Section 4.3.4.4 the slewing of the OTA has to be taken into account if the following condition is fulfilled for the OTA input voltage V_a

$$|V_a| > \frac{I_{SR}}{g_m} = 250\,\text{mV} \tag{4.144}$$

with I_{SR} equal to 11.732 μA. Detailed analysis shows that the reduction in available time for linear settling is negligible in ϕ_1. This is not the case for ϕ_2 where the maximum input voltage at the beginning of the phase equals approximately 1.1 V. The times available for linear settling τ_1 and τ_2 are given by

$$\tau_1 \approx t_s = \frac{1}{2{\cdot}\phi_{sample}} \tag{4.145}$$

$$\tau_2 = t_s - t_{SR,2} = t_s - \frac{\Delta V_{SR,2}}{\frac{I_{SR}}{C_{eq,ol,2}}} = 232\,\mu\text{s} > \frac{1}{2.5{\cdot}\phi_{sample}} \tag{4.146}$$

where the slewing voltage-interval $\Delta V_{SR,2} \approx 1.1$ V.

Noise source	ϕ_1	ϕ_2
R_1	7.04e-14	4.06e-16
R_2	7.04e-14	/
R'_1	2.38e-16 / 7.37e-14	/
R_{BW}	6.61e-14	/
R_{AD}	/	3.80e-16
$\overline{dv^2}_{OP}$	1.35e-14	2.35e-14
Total	**2.94e-13**	**2.43e-14**

Table 4.12: Overview of the amplifier-input-referred total-integrated-noise-powers $N'_{tot,i}$ for the various noise sources in the two phases of the S/H (T=314 K).

In order to obtain in both phases an output-voltage settling-error smaller than 0.1 % of the wanted voltage accuracy-level $V_{err,\sigma}$, δ_1 and δ_2 must now satisfy the following conditions

$$\delta_1 < 0.1\% \cdot \frac{V_{err,\sigma}}{250 \text{ mV}} \tag{4.147}$$

$$\delta_2 < 0.1\% \cdot \frac{V_{err,\sigma}}{1.1 \text{ V}} \tag{4.148}$$

so that the conditions for the closed-loop poles of the implemented S/H become

$$\frac{p_{cl,1}}{\phi_{sample}} > \frac{-\ln(\delta_1) \cdot \rho_1}{2\pi \cdot \tau_1 \cdot \phi_{sample}} \approx 3.7 \tag{4.149}$$

$$\frac{p_{cl,2}}{\phi_{sample}} > \frac{-\ln(\delta_2) \cdot \rho_2}{2\pi \cdot \tau_2 \cdot \phi_{sample}} \approx 5.2 \tag{4.150}$$

For the implemented S/H the ratios $\frac{p_{cl,1}}{\phi_{sample}}$ and $\frac{p_{cl,2}}{\phi_{sample}}$ are equal to 14.2 and 18.7 respectively, satisfying the above conditions. Detailed analysis also shows that the influence on the settling behavior of the on-resistances of the implemented switches and the parasitic capacitance of the $C_{S/H}$ in ϕ_1 is negligible.

4.3.5.4 Noise

In this section the equivalent noise referred to the input of the AMP, resulting from the various noise components of the S/H, is determined. An overview of all the noise sources in the two phases (cf Fig. 4.31) is shown in Fig. 4.35. The noise excess factor γ (cf Eq. (4.102)) of the S/H OTA equals 6 and the maximum switch resistances are equal to 1500 Ω. The same techniques as in Section 4.3.4.7 are used to determine the amplifier-input-referred noise in the two phases. The results are shown in Table 4.12. Note that R'_1 gives rise to a noise contribution at the negative (left) and positive (right) OTA side. The total amplifier-input-referred rms noise voltage is equal to 564 nV, so that the total amplifier-input-referred noise voltage due to the AMP and the S/H equals 8.84 μV $\approx 0.27 \cdot V_{err,\sigma,in}$($\approx$ 32.9 μV).

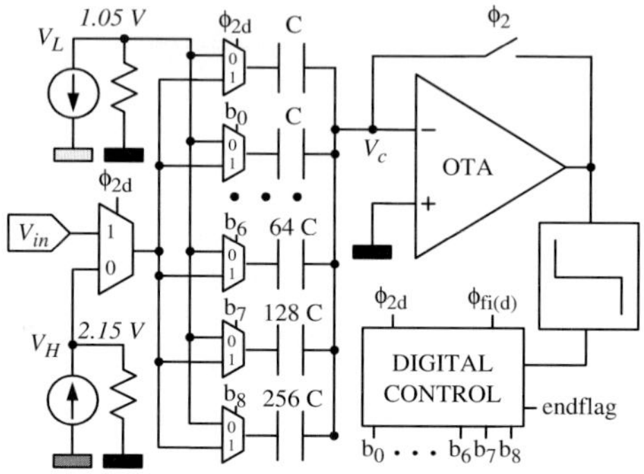

Figure 4.36: *Successive approximation ADC.*

4.3.6 ADC

4.3.6.1 Operating principle

The next building block is the analog-to-digital converter ADC (cf Fig. 4.19). In this section the final ADC, implemented in the intelligent-datalogger IC (cf Chapter 5), is explained. The ADC is a 9-bit successive approximation ADC. This kind of converter applies a binary search algorithm to determine the closest digital word to match the input signal. The main building blocks of the ADC, shown in Fig. 4.36, are a charge-redistribution DAC [McC 75, Sua 75], a comparator with an offset-cancelled preamplifier (OTA), and a digital control unit. The latter performs the binary search sequentially, i.e. in every conversion cycle one single bit is determined. In Fig. 4.36 V_{MM} is assumed grounded (cf Section 4.3.4.1).

Fig. 4.37 illustrates the complete conversion process [Joh 97] of the ADC. It operates as follows:

- **1.** In the reset phase, when ϕ_2 is high, all the capacitors of the charge-redistribution DAC are charged to V_{in}-V_{off} by setting all the bits b_i high. V_{off} is the offset voltage of the comparator's preamplifier, which is cancelled in the remainder of the conversion process.
- **2.** Next, when ϕ_2 goes low, the preamplifier of the comparator is taken out of the reset state by opening the switch connected to its virtual ground, and all the capacitors with exception of the MSB capacitor (i.e. 256 C) are switched to the low reference voltage V_L by setting $b_{7..0}$ low. The MSB capacitor is switched to the high reference voltage V_H. This causes the voltage at the input of the comparator V_c to change to V_L-V_{in}+$V_{ref}/2$ with V_{ref}=V_H-V_L=1.1 V. The voltage V_L-V_{in} lies within the interval between -1.1 V and 0 V (cf Fig. 4.10). Note that a capacitor with size C (upper capacitor in Fig. 4.36) is added to get exact divisions by powers of two. The applied voltage step to the input of the comparator in the different conversion steps is given by

$$\frac{2^i \cdot C}{C_{tot}} \cdot V_{ref} = \frac{1}{2^{9-i}} \cdot V_{ref} \qquad (4.151)$$

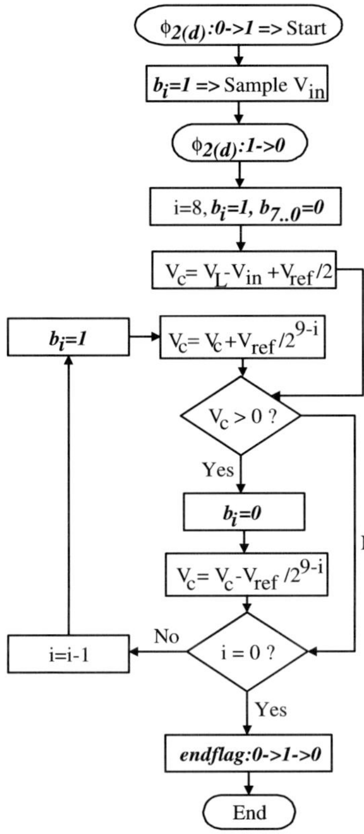

Figure 4.37: *Flowchart of the conversion process.* V_{ref}*=1.1 V.*

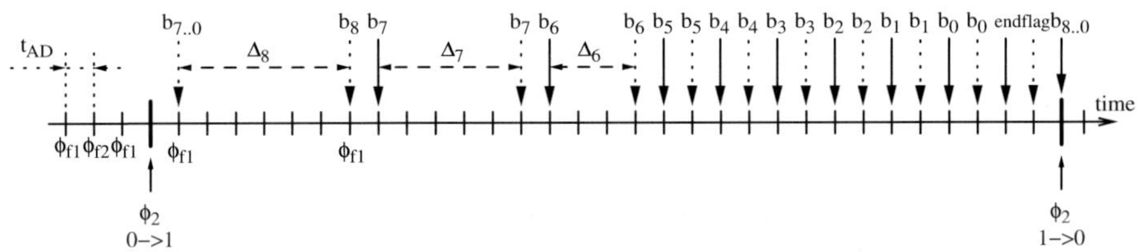

Figure 4.38: *Timing of the AD conversion.*

- **3.** Next, the comparator input voltage V_c is compared with the ground, i.e. 0 V . If V_c is negative, the MSB capacitor is left connected to V_H and b_8 is considered to be a 1. Otherwise, the MSB capacitor is connected to V_L by setting b_8 low, so that the voltage at the input of the comparator V_c changes with $-V_{ref}/2$. This process is repeated with a smaller capacitor being switched each time, until the conversion is finished. The resulting voltage V_c at the end of the conversion is within $\pm$ 1/2 LSB ($=\frac{1.1\ V}{512}$) of the ground and the output of the digital unit $b_8b_7b_6b_5b_4b_3b_2b_1b_0$ represents V_{in}.

The ADC is clocked by the non-overlapping clocks $\phi_{fi\ (d)}$. These clocks have a period of 64 kHz. Fig. 4.38 shows the timing of the conversion process. One single time step t_{AD} in Fig. 4.38 corresponds with half a period of $\phi_{fi\ (d)}$, i.e. 7.8 μs. The arrows with a solid line indicate when the bits b_i are set high. This is done when ϕ_{f2} goes high. On the other hand, the arrows with a dotted line indicate when the comparisons are carried out, occurring when ϕ_{f1} goes high. At these moments it is decided whether b_i is a 1 or a 0. Note that the conversion times Δ_i between the setting of the bits and the comparisons are longer for the 3 MSB bits than for the other bits. For the 3 MSB bits Δ_8, Δ_7 and Δ_6 are equal to $6t_{AD}$, $5t_{AD}$ and $3t_{AD}$ respectively, while for the other bits Δ_i equals t_{AD}. This is done to ease the settling specification (cf Section 4.3.6.5) for the MSB bits where the largest voltage steps ($=\frac{1}{2^{9-i}}\cdot V_{ref}$) occur. When the conversion is carried out, this is notified to the digital part of the datalogger by the endflag-bit.

The realization of the successive approximation algorithm is illustrated in Fig. 4.39. The implementations of the D-flipflop and the SR-latch, used in Fig. 4.39, are given in Fig. 4.40 [Rab 96].

4.3.6.2 Charge redistribution DAC

The accuracy of the ADC is determined by the accuracy of the charge-redistribution DAC, which is a binary-weighted capacitor array. A similar approach as in Section 4.3.2.2 is followed to determine the required accuracy of the unit capacitors C forming the capacitor array.

The DAC is designed with an INL-specification of 0.1 LSB. First, the INL-requirement due to *mismatch* is considered. This is taken equal to 0.08 LSB, so that the total-INL-specification of 0.1 LSB, including also a systematic INL-error (cf below), is satisfied. The required accuracy for the unit capacitors σ_C/C for a given INL-requirement and a wanted yield for a 9-bit resolution binary-weighted DAC is given by (cf [Van 00])

$$\frac{\sigma_C}{C} \leq \frac{\text{INL}}{\sqrt{2^9}\cdot C_{yield}} \tag{4.152}$$

where

$$C_{Yield} = \text{invnorm}_{(-\infty,x)}(0.75 + \frac{\text{yield}}{4}) \tag{4.153}$$

with $\text{invnorm}_{(-\infty,x)}$ the inverse function of the normal cumulative function integrated from $-\infty$ to x. Fig. 4.41 shows the yield for an INL-requirement of 0.08 LSB, predicted by Eq. (4.152) and Eq. (4.153) for different values of σ_C/C. In order to achieve a yield better than 99.7 %, the capacitor mismatch σ_C/C must be smaller than 0.11 %.

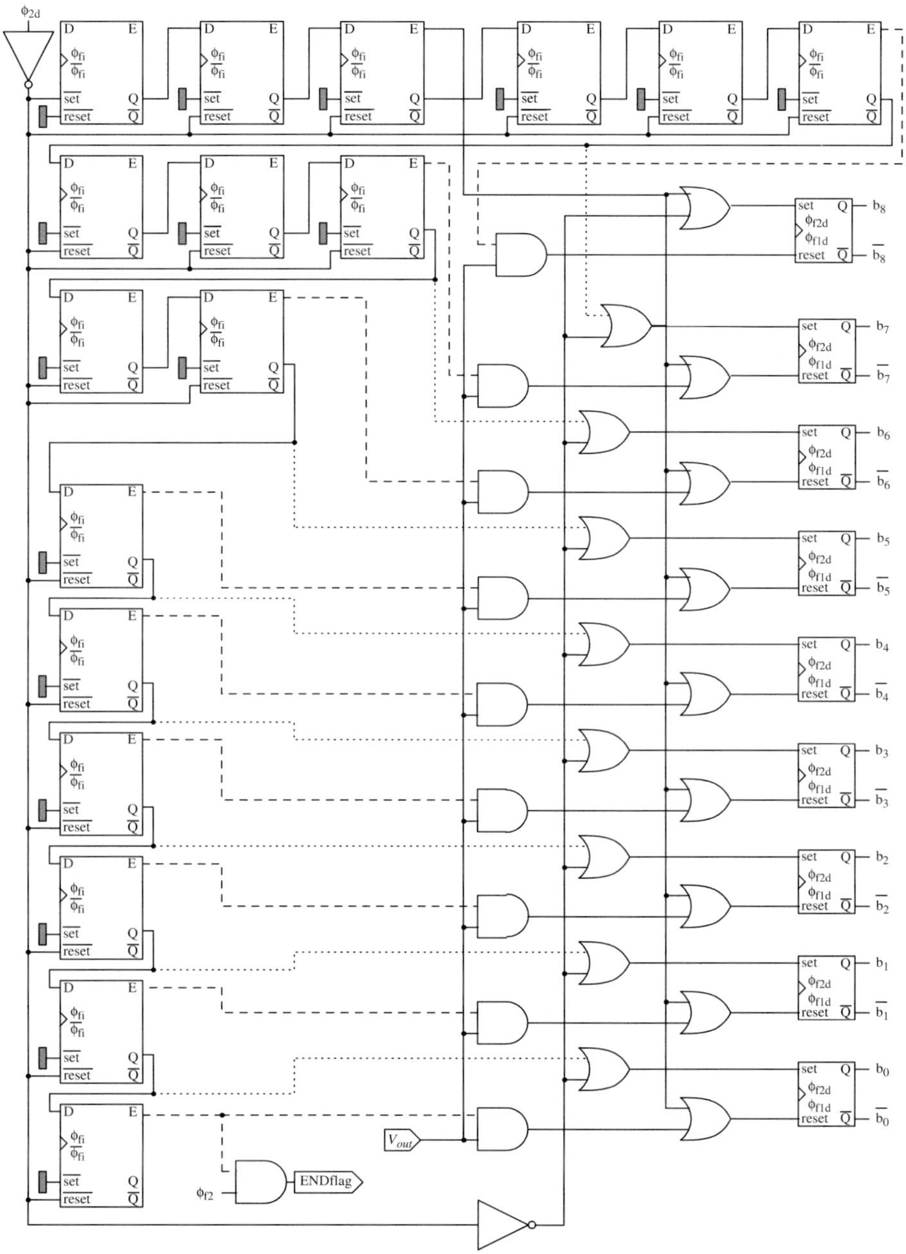

Figure 4.39: *Digital control unit. V_{out} is the comparator output.*

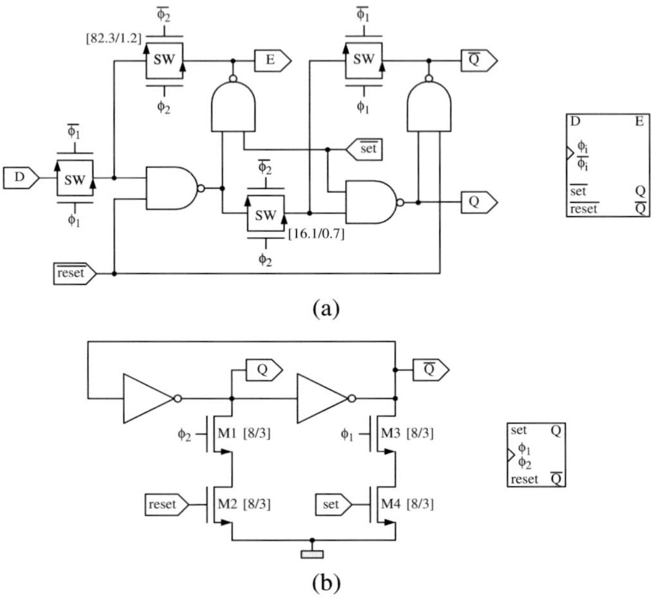

Figure 4.40: *(a) D-flipflop with set and reset. (b) SR-latch.*

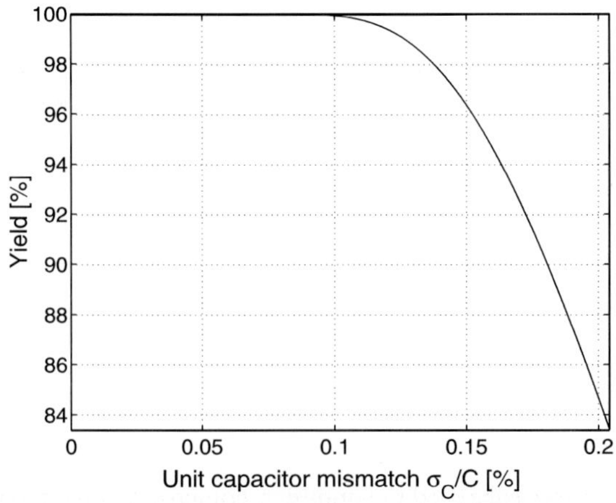

Figure 4.41: *Yield for a 9-bit resolution ADC with a 0.08-LSB INL-specification as a function of* σ_C/C.

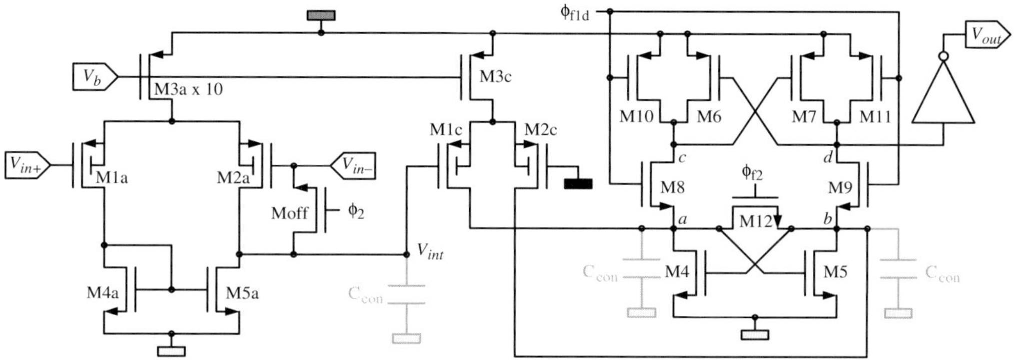

Figure 4.42: *Comparator with offset-cancelled preamplifier (Mia and Moff).* C_{con} *denotes the parasitic capacitance of the connections. Transistor dimensions are listed in Table A.5.*

For the C07MA technology the mismatch between two square unit capacitors C with an area A_{cap} is given by

$$\frac{\sigma_C}{C} = \sqrt{114.17 \cdot A_{cap}^{-3/2} - \frac{1.04}{A_{cap}} + 8.96E^{-8} \cdot A_{cap} - 3.13E^{-4}} \quad [\%] \tag{4.154}$$

From Eq. (4.154) follows that the accuracy of the implemented unit capacitors, which have an area of 400 μm^2 and a capacitance value of 300 fF, equals 0.1067 %, showing that the accuracy condition for the unit capacitors is satisfied.

In order to cope with graded (linear) *systematic* errors the capacitors are placed in a common-centroid layout. Two of the capacitors are left uncompensated by this common-centroid layout, i.e. the two capacitors with size C. The resulting systematic INL-error equals $2 \cdot \frac{E_{p-p}}{2} = E_{p-p}$ with E_{p-p} the worst case peak-to-peak error [Bas 98] given by

$$E_{p-p} = D \cdot 1.58E^{-4} + 0.037 \quad [\%\ \mathrm{LSB}] \tag{4.155}$$

for the C07MA technology. This results in a systematic INL-error of 0.003 LSB, corresponding with a maximum distance D of 1670 μm. The resulting total INL-error equals approximately 0.083 LSB, satisfying the INL-requirement.

4.3.6.3 Comparator

The schematics of the offset-cancelled preamplifier, composed of the transistors Mia and Moff, and the comparator are shown in Fig. 4.42. The comparator circuit [Yin 92, Mar 99, Gee 01] consists of a differential input pair (M1c/M2c), a top and bottom regeneration loop (M6/M7 and M4/M5) with transfer transistors (M8/M9) and pre-charge transistors (M10/M11), and a switch for resetting (M12). The operation of this circuit is as follows. During the clock phase ϕ_{f2}

the comparator is in its reset phase. The top and bottom regeneration loops are disconnected, since the transfer transistors (M8/M9) are off. Nodes *c* and *d* are reset to the power supply V_{dd} by the pre-charge transistors (M10/M11) and the bottom regeneration loop is reset by M12. The differential pair (M1c/M2c) injects a differential current, proportional to the comparator input voltage difference V_{in}, into the bottom regeneration loop and generates a voltage difference between nodes *a* and *b*. This voltage will act as the initial imbalance for the regeneration. When ϕ_{f2} goes down, the imbalance voltage is regenerated by the bottom regeneration loop until ϕ_{f1d} rises. Then, the bottom and the top regeneration loops are connected and they both start to regenerate the imbalance.

The offset of the comparator is determined by the input pair and the bottom regeneration loop. Mismatches in the top regeneration loop can be neglected for the following reason. When ϕ_{f2} goes down at the end of the reset phase, the bottom loop starts to regenerate the initial imbalance between nodes *a* and *b*. The regeneration process is determined by the following equation [Mar 99]

$$V_{ab}(t_{reg}) \approx V_{ab}(0) \cdot \exp\left(\frac{g_{m4}}{C_a} \cdot t_{reg}\right) \tag{4.156}$$

with

$$C_a = C_{dtotM1c} + C_{dtotM12} + C_{gtotM5} + C_{dtotM4} + C_{dtotM8} + C_{con} \approx 340 \text{ fF} \tag{4.157}$$

the capacitance at node *a*, C_{con} the parasitic capacitance of the connections, t_{reg} ($\approx$ 450 ns) the time between the falling edge of ϕ_{f2} and the rising edge of ϕ_{f1d}, $g_{m4} \approx 22 \frac{\mu A}{V}$, and $V_{ab}(0)$ the voltage at the start of the regeneration. Because of the small time constant $\frac{C_a}{g_{m4}}$ in comparison with t_{reg}, the initial imbalance is already significantly larger at the moment when the PMOS loop is connected. Therefore, the offset and clock feedthrough of the top loop can be neglected. When ϕ_{f2} goes down, M12 also introduces an offset error due to charge injection. To restrict this error, M12 is kept small and the comparator is laid out symmetrically, so that the offset error due this switch is also negligible. The resulting amplifier-input-referred offset of the comparator is equal to [Yin 92]

$$\sigma_{offset} \approx \frac{\sqrt{\sigma^2_{M1c/2c} + \frac{g^2_{m4}}{g^2_{m1c}} \cdot \sigma^2_{M4/5}}}{A_{pre} \cdot A_{AMP}} \approx 0.13\ \mu V \tag{4.158}$$

where σ^2_{Mi} is given by Eq. (4.55) and A_{pre} ($\approx$ 300) is the gain of the preamplifier (cf Section 4.3.6.5).

When a very small signal is applied to the comparator, the initial imbalance will be very small and nodes *c* and *d* will start to drop together from the supply voltage towards the meta-stable point of the latch, before the regeneration starts. If the meta-stable point is too low, this might inadvertently trigger the SR-latches which set the bits b_i (cf Fig. 4.39). To avoid improper triggering the meta-stable point is set above the threshold of the output inverter (Fig. 4.42), so that the output voltage V_{out} remains 0 V as long as the imbalance is not regenerated, and the SR-latch is not triggered.

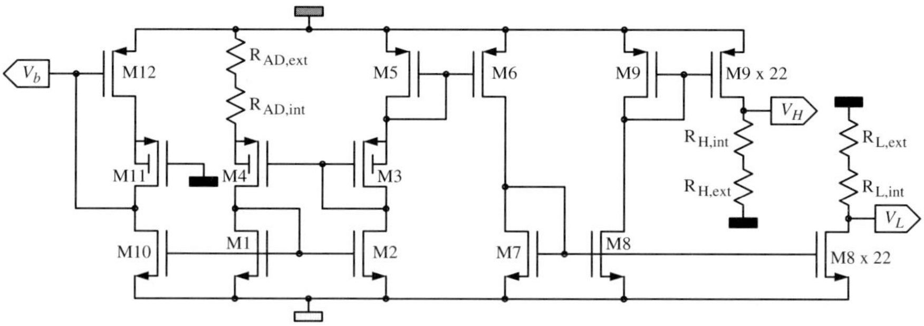

Figure 4.43: *Reference current source of the ADC and derivation of the reference voltages. Start-up circuit is not shown. Transistor dimensions are listed in Table A.6.*

To restrict hysteresis the voltage difference between the voltages at node *a* and *b* at the end of the reset phase due to the previous output in the regeneration phase has to be strongly reduced. The difference between these voltages V_{ab} at the end of the reset phase is given by [Mar 99]

$$V_{ab}(t_{AD}) \approx V_{ab}(0)\cdot\exp\left(\frac{g_{eq}}{C_a}\cdot t_{AD}\right) - \frac{g_{m1c}}{g_{eq}}\cdot V_{in}\cdot\left(1-\exp\left(\frac{g_{eq}}{C_a}\cdot t_{AD}\right)\right) \tag{4.159}$$

with

$$g_{eq} = g_{m4} - 2\cdot g_{ds12} - g_{dsM1c} = -247.2\,\frac{\mu A}{V} \tag{4.160}$$

and $V_{ab}(0)$ (< 3.1 V) the voltage difference at the beginning of the reset phase due to the previous output (after a very short large-signal-reset period [Yin 92, Mar 99]). Because of the small time constant in comparison with t_{AD}, the exponential terms in Eq. (4.159) are negligible and the voltage difference V_{ab} settles almost completely to $-\frac{g_{m1c}}{g_{eq}}\cdot V_{in}$, so that the hysteresis error due to the previous output can be ignored.

4.3.6.4 Reference current source

To ensure that the switching activity of the ADC does not disturb the clean analog power supplies of the conditioning-and-amplification circuits described above, separate power supplies are used for the ADC. Moreover, a separate reference current source is implemented from which the currents of the ADC and the reference voltages are derived. The schematic of this reference current source is shown in Fig. 4.43. The transistors M1-M4 ensure that the source voltages of M3/M4 are identical (neglecting channel-length modulation), so that the voltage across the resistor R_{AD} ($=R_{AD,int}+R_{AD,ext}$), and the gate-source voltage of M5 are the same, which can be expressed as

$$R_{AD}\cdot I_{AD} \approx V_{GS5} \approx V_{T5} + \sqrt{\frac{2\cdot I_{AD}}{\beta_5}} \quad (\approx 949\text{mV}) \tag{4.161}$$

From this equation an expression for the reference current I_{AD} can be derived [Opt 86]

$$I_{AD} \approx \frac{1 + \beta_5 \cdot V_{T5} \cdot R_{AD} + \sqrt{1 + 2 \cdot \beta_5 \cdot V_{T5} \cdot R_{AD}}}{\beta_5 \cdot R_{AD}^2} \tag{4.162}$$

Note that the reference-current circuit is also stable if all the transistors carry a zero current. To solve this problem a start-up circuit (not shown in Fig. 4.43), that drives the circuit out of its degenerated bias point, is added. The reference current I_{AD} is equal to 5 μA.

To cope with the limited absolute accuracy of the on-chip high-ohmic poly resistors, i.e. ± 20 % [Alc 01], R_{AD} is implemented with an on-chip high-ohmic poly resistor $R_{AD,int}$ and an off-chip laser-trimmable screen-printed resistor $R_{AD,ext}$. The values of these resistors are given in Table 4.13.

From Eq. (4.161) the temperature dependence of I_{AD} can be derived [Opt 86]

$$\frac{1}{I_{AD}} \cdot \frac{\delta I_{AD}}{\delta T} = \frac{\frac{\delta V_{T5}}{\delta T} - 0.5 \cdot \sqrt{\frac{2 \cdot I_{AD}}{\beta_5}} \cdot \frac{1}{\beta_5} \cdot \frac{\delta \beta_5}{\delta T} - I_{AD} \cdot \frac{\delta R_{AD}}{\delta T}}{R_{AD} \cdot I_{AD} - 0.5 \cdot \sqrt{\frac{2 \cdot I_{AD}}{\beta_5}}} \tag{4.163}$$

with [Alc 01]

$$\frac{\delta V_{T5}}{\delta T} = -1.9 \ \frac{mV}{^{o}C} \tag{4.164}$$

$$\frac{1}{\beta_5} \cdot \frac{\delta \beta_5}{\delta T} = -\frac{n}{T} = -\frac{1.47}{T} \tag{4.165}$$

$$\begin{aligned} \delta R_{AD} = \delta R_{AD,int} + \delta R_{AD,ext} = R_{AD,int} \cdot TCL_{int} \cdot \delta T + R_{AD,int} \cdot TCQ_{int} \\ \cdot [(t_{ref} + \delta T - 30\ ^{o}C)^2 - (t_{ref} - 30\ ^{o}C)^2] + R_{AD,ext} \cdot TCL_{ext} \cdot \delta T \end{aligned} \tag{4.166}$$

In Eq. (4.166) TCL_{int},TCQ_{int} and TCL_{ext} are respectively the linear temperature coefficient and the quadratic temperature coefficient of $R_{AD,int}$, and the linear temperature coefficient of $R_{AD,ext}$. The values of these temperature coefficients are also given in Table 4.13.

The reference voltages are equal to V_H=$R_H \cdot I'_{AD}$=0.6 V and V_L=-$R_L \cdot I'_{AD}$=-0.5 V with I'_{AD}=22·I_{AD}. Each of the reference resistors $R_{H/L}$ consists of an on-chip low-ohmic poly resistor, having an absolute accuracy of ± 26 % [Alc 01], and an off-chip laser-trimmable screen-printed resistor (cf Fig. 4.43). The temperature dependence of the reference voltages V_H and V_L can be expressed as

$$\begin{aligned} \pm \delta V_{H/L} = \left[R_{H/L,int} + R_{H/L,ext}\right] \cdot \frac{\delta I'_{AD}}{\delta T} \cdot \delta T + I'_{AD} \cdot \left[R_{H/L,int} \cdot TCL'_{int} \cdot \delta T + R_{H/L,int} \cdot TCQ'_{int} \right. \\ \left. \cdot [(t_{ref} + \delta T - 30\ ^{o}C)^2 - (t_{ref} - 30\ ^{o}C)^2] + R_{H/L,ext} \cdot TCL'_{ext} \cdot \delta T \right] \end{aligned} \tag{4.167}$$

where the negative sign is valid for δV_L and the positive for δV_H. The temperature coefficients are now denoted by accents (cf Table 4.13). In order to obtain a low temperature dependence a resistor paste with a negative linear temperature coefficient of -1700 ppm/^{o}C [Ele] is employed for the screen-printed resistors of the reference voltages.

Resistor	Value
$R_{AD,int}$	$117.2 \pm 20\%$ kΩ
$R_{AD,ext}$	72.6 kΩ
$R_{H,int}$	$4120 \pm 26\%$ Ω
$R_{H,ext}$	1101 Ω
$R_{L,int}$	$3434 \pm 26\%$ Ω
$R_{L,ext}$	1301 Ω

Temp. coeff.	Value
TCL_{int}	-2100 ppm/ °C
TCQ_{int}	5 ppm/ °C^2
TCL'_{int}	620 ppm/ °C
TCQ'_{int}	1.1 ppm/ °C^2
TCL_{ext}	< 50 ppm/ °C
TCL'_{ext}	-1700 ppm/ °C

Table 4.13: Overview of the resistor values (at 36.4 °C) and the temperature coefficients [Dup b, Ele].

The resulting temperature-induced error voltage at the input of the comparator is given by (cf Fig. 4.37)

$$\delta V_c = \delta V_L + \frac{j}{2^9} \cdot (\delta V_H - \delta V_L) \tag{4.168}$$

with j the digital output ($0 \leq j \leq 511$). From Eq. (4.163), Eq. (4.167) and Eq. (4.168) follows that for the nominal resistor values the maximum temperature-induced error over the oral-temperature-interval of 0.5-0.7 °C [Gan 95] equals 0.02 LSB. This corresponds with the simulated value of 0.02 LSB. The simulated worst-case mean error of V_c, corresponding with a maximum deviation of all the on-chip resistors, equals 0.01 LSB.

The power-supply dependence of I_{AD} can be found by small-signal analysis and is given by

$$\frac{\delta I_{AD}}{\delta V_{dd,ss}} = \frac{\frac{g_{o2}+g_{o4}}{g_{m4}} + R_{AD} \cdot g_{o2}}{R_{AD} - \frac{1}{g_{m5}}} \tag{4.169}$$

This results in a power-supply dependence of -142 dB, which is consistent with the simulated value of -142.1 dB. The maximum DC variation from 3.1 V of the operating voltage is limited to ± 40 mV, which gives rise to a worst-case mean error of V_c equal to 0.06 LSB.

The noise power of the reference current I_{AD} is also found by small-signal analysis. An approximated expression is given by

$$\overline{di^2}_{I_{AD}} \approx \overline{di^2}_{M1} + \overline{di^2}_{M2} + \left(\frac{R_{AD} - r_{o2}}{g_{m4} \cdot R_{AD} \cdot r_{o2}}\right)^2 \cdot \overline{di^2}_{M3} + \left(\frac{1}{g_{m4} \cdot R_{AD}}\right)^2 \cdot \overline{di^2}_{M4} + \overline{di^2}_{M5} + \overline{di^2}_{R_{AD}} \tag{4.170}$$

A noise power of $0.15 \cdot 10^{-23}\ \frac{A^2}{Hz}$ is found with Eq. (4.170) which corresponds well with the value found if no simplifications are made in the small-signal analysis, i.e. $0.22 \cdot 10^{-23}\ \frac{A^2}{Hz}$. This value will be used further in the noise calculations (cf Section 4.3.6.6).

4.3.6.5 Settling behavior

In this section the settling behavior of the ADC is investigated. First the *reset phase* ϕ_2, shown in Fig. 4.44 (a), is considered. The output-voltage error is given by Eq. (4.60). ΔV_{ϕ_2} is maximally

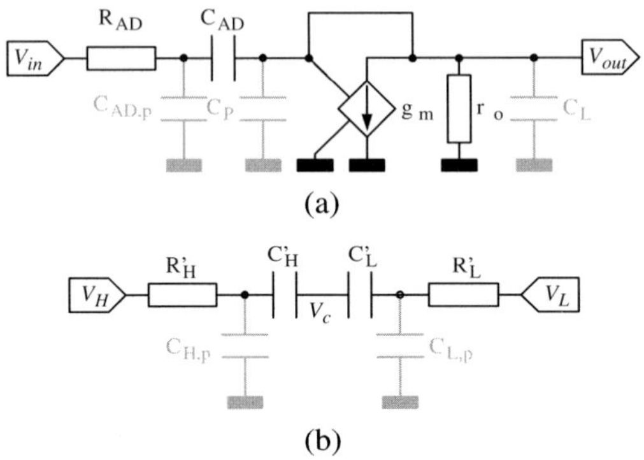

g_m	200 μA/V
r_o	1.5 MΩ
I_{SR}	50 μA
C_P	835 fF
C_L	520 fF
C_{AD}	153.6 pF
R_{AD}	1.5 kΩ
R'_H	6221 Ω
R'_L	5234 Ω

Figure 4.44: *The ADC (a) during the reset phase ϕ_2 and (b) during the conversion phase ϕ_1.*

equal to 250 mV, which is found by charge-conservation analysis, and δ_2 is equal to

$$\delta_2 = \exp\left(-\frac{p_{cl,2}}{1+\lambda_2}\cdot\frac{\tau_2}{\rho_2}\right) \tag{4.171}$$

where

$$\lambda_2 = p_{cl,2}\cdot R_{AD}\cdot C_{AD} \tag{4.172}$$

models the influence of the resistance R_{AD} on the settling behavior. Note that the influence of the parasitic capacitance $C_{AD,p}$ of C_{AD} on the settling behavior in ϕ_2 is negligible. The closed-loop pole $p_{cl,2}$ is given by Eq. (4.82) and Eq. (4.86) where the equivalent open-loop capacitance now equals

$$C_{eq,ol,2} = C_L + C_P + C_{AD} \tag{4.173}$$

and the feedback factor f_{dc2} is equal to 1. Detailed analysis shows that for the implemented amplifier the reduction in available time for linear settling due to slewing is negligible in ϕ_2, so that τ_2 is approximately given by Eq. (4.91).

In order to obtain an output-voltage settling-error smaller than 0.1 % of the wanted voltage accuracy-level $V_{err,\sigma}$, δ_2 must satisfy the following condition

$$\delta_2 < 0.1\%\cdot\frac{V_{err,\sigma}}{250\text{ mV}} \tag{4.174}$$

so that the condition for the closed-loop pole $p_{cl,2}$ of the implemented ADC during the reset phase becomes

$$\frac{p'_{cl,2}}{\phi_{sample}} = \frac{\frac{p_{cl,2}}{1+p_{cl,2}\cdot R_{AD}\cdot C_{AD}}}{\phi_{sample}} > \frac{-\ln(\delta_2)\cdot\rho_2}{2\pi\cdot\tau_2\cdot\phi_{sample}} \approx 3.7 \tag{4.175}$$

For the implemented ADC the ratio $\frac{p'_{cl,2}}{\phi_{sample}}$ is equal to 79.2, satisfying the above condition.

Next the settling condition of the OTA in the *conversion phase* ϕ_1 is discussed. In order to obtain an equivalent OTA-output-voltage settling-error smaller than 0.1 % of the wanted voltage accuracy-level $V_{err,\sigma}$ in ϕ_1, δ_1 must satisfy the following condition

$$\delta_1 < 0.1\% \cdot \frac{A_{pre} \cdot V_{err,\sigma}}{\Delta V_{\phi_1}} \tag{4.176}$$

where A_{pre} is the gain (≈ 300) of the OTA and ΔV_{ϕ_1} equals at most 1.55 V when approaching the comparator's threshold (i.e. 1.55 V). From this follows the condition for the open-loop pole $p_{ol,1}$ which is given by

$$\frac{p_{ol,1}}{\phi_{sample}} > \frac{-\ln(\delta_1)}{2\pi \cdot \tau_1 \cdot \phi_{sample}} \approx 78.6 \tag{4.177}$$

with $\tau_1 = t_{AD} = \frac{1}{64 \cdot \phi_{sample}}$. The open-loop pole $p_{ol,1}$ can be expressed as

$$p_{ol,1} = \frac{g_m}{2\pi \cdot A_{pre} \cdot C_L} \quad [\text{Hz}] \tag{4.178}$$

which is equal to $102 \cdot \phi_{sample}$, satisfying Eq. (4.175).

In the *conversion phase* the charge-redistribution DAC also gives rise to settling errors. Fig. 4.44 (b) shows the equivalent schematic of the DAC in ϕ_1. Here R'_H is the sum of the reference resistor R_H and the maximum resistance (<2 x 0.5 kΩ) of the switches (cf Fig. 4.36). C'_H is the total capacitance composed of unit capacitances connected to V_H and is equal to $j \cdot C$ with j the digital output code which changes during the conversion process. $C_{H,p}$ is the parasitic capacitance of C_H with a maximum value of $0.25 \cdot C_H + C_{H,con}$. The latter is the parasitic capacitance due to the connections to the unit capacitances and is overestimated at 20 pF. The same definitions are valid for V_L where C'_L is equal to $(512 - j) \cdot C$.

To investigate the settling behavior of this circuit the poles are determined by applying Kirchoff's laws. They are given by

$$p_1 = \frac{1}{\tau_1} = \frac{B + \sqrt{B^2 - 4 \cdot A}}{2 \cdot A} \tag{4.179}$$

$$p_2 = \frac{1}{\tau_2} = \frac{B - \sqrt{B^2 - 4 \cdot A}}{2 \cdot A} \tag{4.180}$$

with

$$A = R'_H \cdot R'_L \cdot [C_{H/L} \cdot (C_{H,p} + C_{L,p}) + C_{H,p} \cdot C_{L,p}] \tag{4.181}$$

$$B = C_{H/L} \cdot (R'_H + R'_L) + C_{H,p} \cdot R'_H + C_{L,p} \cdot R'_L \tag{4.182}$$

where $C_{H/L}$ is the equivalent capacitance of the series connection of C_H and C_L:

$$C_{H/L} = \frac{j \cdot (512 - j)}{512} \cdot C \tag{4.183}$$

i	Δ_i	$V_{err,\,\phi_{1,i}}$
8	$6t_{AD}$	≈ 0
7	$5t_{AD}$	≈ 0
6	$3t_{AD}$	≈ 0
5	t_{AD}	$0.037\% \cdot V_{err,\,\sigma}$
4	t_{AD}	$0.018\% \cdot V_{err,\,\sigma}$
3	t_{AD}	$0.009\% \cdot V_{err,\,\sigma}$

Table 4.14: Conversion times and maximum settling errors for the MSB conversions.

The most stringent poles are found for j=256 when the maximum of $C_{H/L}$ is reached. For this condition p_1 and p_2 are given by $-1.5 \cdot 10^6 \frac{rad}{s}$ and $-4.5 \cdot 10^6 \frac{rad}{s}$.

The resulting error of the second-order system with real poles can be approximated by [Pol 94]

$$V_{err,\phi_{1,i}} \approx \left[-\frac{\tau_1}{\tau_1 - \tau_2} \cdot \exp\left(\frac{-\Delta_i}{\tau_1}\right) + \frac{\tau_2}{\tau_1 - \tau_2} \cdot \exp\left(\frac{-\Delta_i}{\tau_2}\right) \right] \cdot \Delta V_{\phi_{1,i}} \tag{4.184}$$

where the voltage steps $\Delta V_{\phi_{1,i}} = \frac{1}{2^{9-i}} \cdot V_{ref}$ depend on the conversion step (cf Fig. 4.37). As already indicated in Section 4.3.6.1 the conversion times Δ_i are longer for the 3 MSB bits than for the other bits. The resulting maximum settling errors for the MSB conversions are given in Table 4.14, showing that the error is smaller than $0.1\% \cdot V_{err,\sigma}$ for all the conversion steps.

4.3.6.6 Noise

In this section the equivalent noise referred to the input of the amplifier AMP, resulting from the various noise components of the ADC, is investigated. Fig. 4.45 gives an overview of the noise sources in the reset phase ϕ_2 and the conversion phase ϕ_1. In ϕ_1 all the noise sources are shown, except for the noise source due to the OTA. The noise excess factor γ (cf Eq. (4.101)) of the OTA is equal to 2.5 and the maximum switch resistances $R_{H,sw}$ and $R_{L,sw}$ are equal to 1 kΩ and 0.5 kΩ respectively. In the calculations the parasitic capacitances are taken equal to their minimum value, i.e. $0.1 \cdot C'_{H/L}$ (cf Table 4.6). Similar techniques as described in Section 4.3.4.7 are employed to determine the amplifier-input-referred noise.

Note that the noise current of the reference current source (cf Eq. (4.170)) gives rise to noise contributions at both the reference voltages V_H and L_L or in other words this noise source gives rise to correlated noise contributions. This is also the case for the transistors M6, M7 and M8 (cf Fig. 4.43). The fact that these noise sources result in correlated noise contributions has been taken into account in the calculations. Table 4.15 gives the resulting noise voltages referred to the input of the comparator in the two phases. The noise voltages are given for j=0, j=1, j=256 and j=511. Note that for j=0 all the unit capacitances are connected to V_L. The maximum total amplifier-input-referred rms noise voltage due to the ADC is equal to 1.61 μV, so that the total

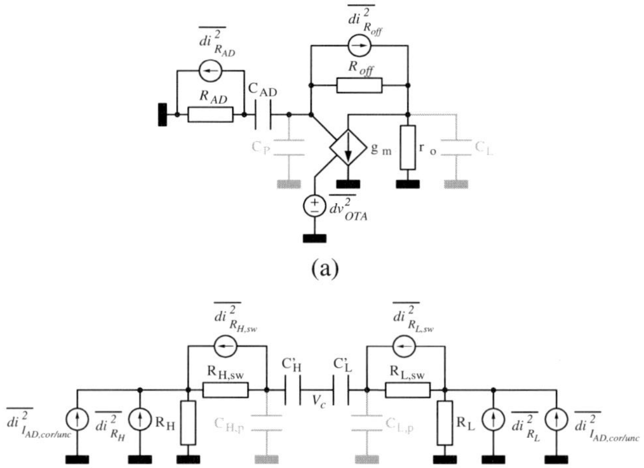

Figure 4.45: *The different noise components (a) during ϕ_2 and (b) during ϕ_1 (OTA is not shown).*

j	ϕ_1	ϕ_2
0	105.9 μV	8 μV
1	99.7 μV	
256	107.8 μV	
511	112.4 μV	

Table 4.15: Overview of the noise voltages referred to the input of the comparator in the two phases.

amplifier-input-referred noise voltage V_{noise} due to the AMP, the S/H and the ADC equals 9 μV $\approx 0.27 \cdot V_{err,\sigma,in}(\approx 32.9\ \mu V)$.

The implemented ADC has a resolution of 9 bits. For an ideal 9-bit ADC, the quantization noise V_{quant} [Joh 97] is given by

$$V_{quant} = \frac{1\ \text{LSB}}{\sqrt{12}} = 0.27 \cdot V_{err,\sigma,in} \tag{4.185}$$

so that the input-referred minimum accuracy level V_{acc} of the total system is approximately given by

$$V_{acc} < \sqrt{V_{noise}^2 + V_{quant}^2} + \text{INL}_{ADC} \approx 0.48 \cdot V_{err,\sigma,in} \tag{4.186}$$

which shows that a safety margin of approximately $0.5 \cdot V_{err,\sigma,in}$ is present.

4.3.7 Oscillator

4.3.7.1 Operating principle and implementation

The last building block of the sensor interface chip is the relaxation clock oscillator CLOCK (cf Fig. 4.19). The schematic of this oscillator is shown in Fig. 4.46 [Wou 95]. Its basic principle is the continuous charging and discharging of a capacitor C_{OSC} by means of constant currents. The voltage across the capacitor is monitored by a Schmitt trigger. The lower and upper thresholds of the Schmitt trigger are given by [Wou 95]

$$V_{S,low} = \frac{\sqrt{\frac{W_1}{L_1}/\frac{W_5}{L_5}} \cdot (V_{DD} + V_{T1}) + V_{T2} - V_{T5}}{1 + \sqrt{\frac{W_1}{L_1}/\frac{W_5}{L_5}}} \quad (4.187)$$

$$V_{S,high} = \frac{\sqrt{\frac{W_4}{L_4}/\frac{W_6}{L_6}} \cdot V_{T4} + V_{T3} - V_{T6} + V_{DD}}{1 + \sqrt{\frac{W_4}{L_4}/\frac{W_6}{L_6}}} \quad (4.188)$$

For the implemented Schmitt trigger $V_{S,high}$ and $V_{S,low}$ are approximately equal to 2.227 V and 853 mV. The output of the Schmitt trigger decides whether the capacitor is being charged or discharged. It controls the switches SW_2 which connect the capacitor to the charging current source, composed of PMOS transistors (Mp), or the discharging current source, composed of NMOS transistors (Mn). Both these current sources consist of 6 binary-weighted current mirrors, which can be (dis)connected to (from) C_{OSC} by means of the switches SW_1, which are controlled by the digital word $c_5c_4c_3c_2c_1c_0$. One current mirror of each current source (Mnx4/Mpx4) is connected at all times, ensuring a basic operation frequency even when all digital controls are set to zero. The reference voltages of the current mirrors V_{bn} and V_{bp} are derived from a reference current source of which the schematic is illustrated in Fig. 4.47. It has the same topology as the reference current source of the ADC (Section 4.3.6.4). The reference current I_{CL} now equals 200 nA. The (dis)charging current consists of a basic component of 800 nA and an adjustable component between 50 nA and 50+100+200+400+800+1600 nA. This battery of current mirrors guarantees that an oscillator frequency close to 128 kHz can be obtained for each process type (slow-typical-fast), taking also into account the tolerance of the oscillator capacitance C_{OSC}, i.e. 13.33 %. Note that a power-on-reset circuit (not shown) is implemented to switch on the oscillator if the sensor interface chip is connected to its power supply. The oscillator is switched on and off by the res-bit (Fig. 4.46).

The clock period T_{osc} of the oscillator is inversely proportional to the (dis)charging current and proportional to the voltage difference between the lower and the upper threshold of the Schmitt trigger. By combining Eq. (4.163), Eq. (4.164), Eq. (4.169), Eq. (4.187) and Eq. (4.188) the accuracy of the clock frequency over the supply-voltage interval (i.e. ± 1 %) and the operating-temperature interval (i.e. ± 0.03 %) of the datalogger can be calculated. These values correspond well with the simulated values of ± 1.2 % and ± 0.05 %. Note that for an NMOS transistor in Eq. (4.187) and Eq. (4.188) the temperature dependence of the threshold voltage $\delta V_{T,n}/\delta T$ is

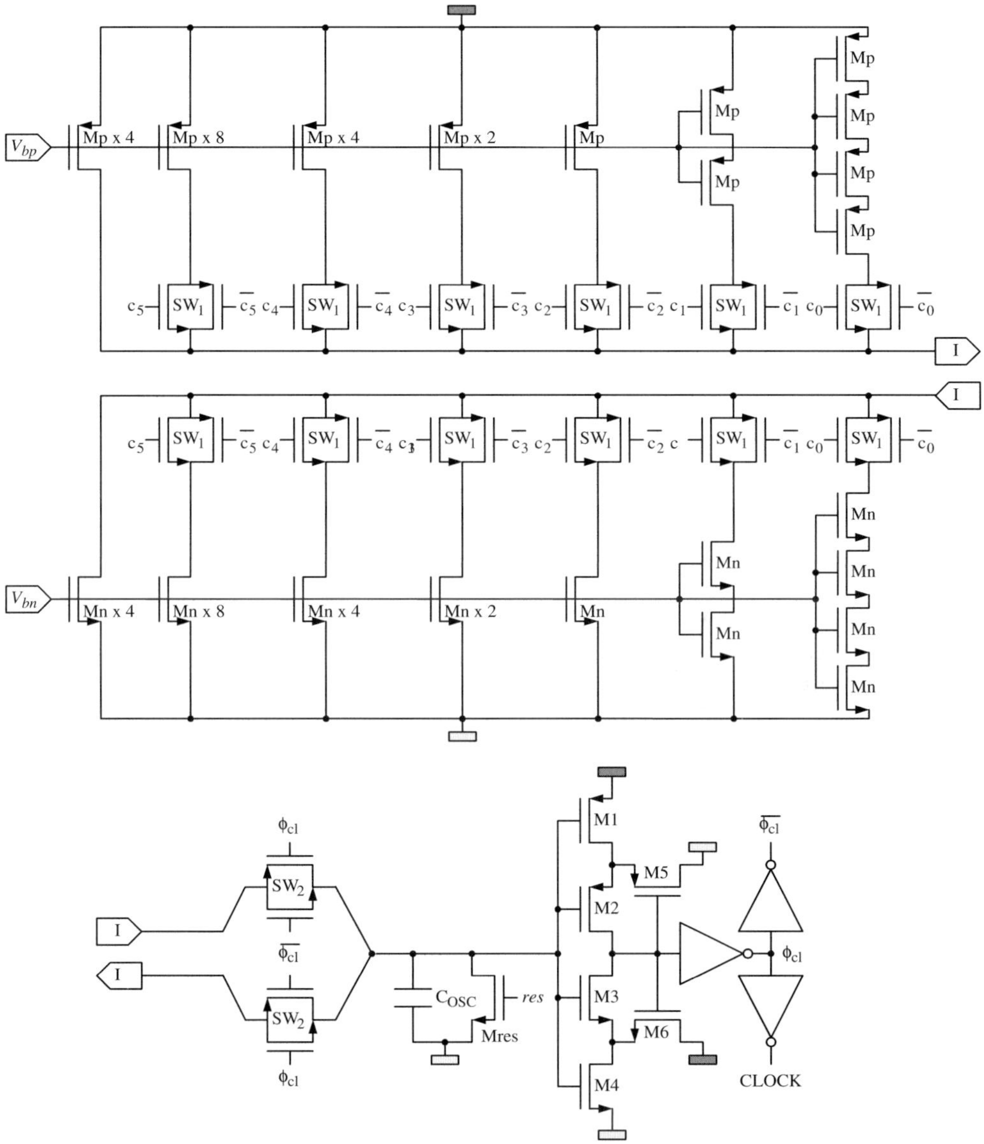

Figure 4.46: *6-bit programmable relaxation oscillator (cf Table A.7).*

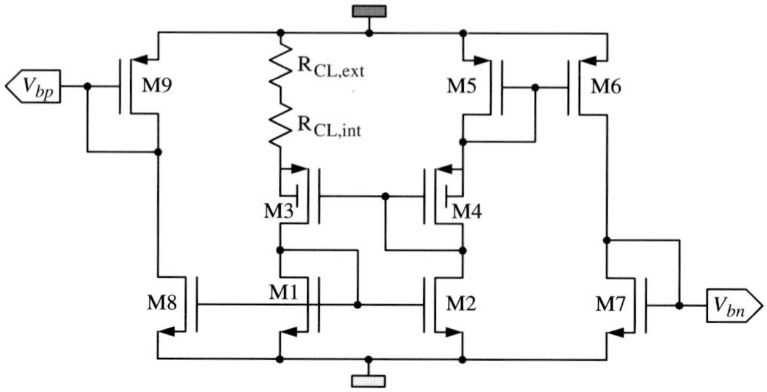

Figure 4.47: *Reference current source of the relaxation clock oscillator. Start-up circuit is not shown. Dimensions/values are given in Table A.8.*

given by

$$\frac{\delta V_{T,n}}{\delta T} = -1.4\,\frac{mV}{^{o}C} \tag{4.189}$$

In [Abi 83] a method to estimate the jitter $\sigma(\delta T)$ of a relaxation clock oscillator is proposed. This approximation is given by

$$\sigma(\delta T) \approx \alpha\cdot\sqrt{2\cdot\frac{V^2_{n,S,low}}{S^2} + 4\cdot\frac{V^2_{n,S,high}}{S^2}} \tag{4.190}$$

where S is the slope of the waveform of the capacitor voltage at the triggering points, $V_{n,S,low}$ and $V_{n,S,high}$ are the rms noise voltages at the input of the Schmitt trigger, when the input voltage approaches the triggering points, and α is a constant, dependent on the rms noise voltage, the noise bandwidth and the capacitor voltage slope [Abi 83]. For the implemented oscillator the calculated fractional jitter $\sigma(\delta T)/T_{osc}$ is approximately equal to 107 ppm, so that the maximum error due to jitter

$$\sigma_{jitter,max} = 2\pi\cdot f_{max}\cdot A_{max}\cdot\sigma(\delta T) \tag{4.191}$$

corresponds with an error of 0.03 μstrain at the maximum frequency f_{max} of 50 Hz and the maximum amplitude A_{max} corresponding with the maximum/minimum strain $\varepsilon_{max,min}$ of ± 1948 μstrain.

4.3.7.2 Non-overlapping clock generators and ϕ_{sample}

To derive the 2-kHz and 64-kHz system clocks from the 128-kHz relaxation clock oscillator, the circuit of Fig. 4.48 is implemented. The 128-kHz clock is first divided by 2 by one D-flipflop, and further divided by 32 by 5 successive D-flipflops. The bi-phasic (delayed) non-overlapping clocks $\phi_{i(d)}$ and $\phi_{fi\,(d)}$, required for the AMP, S/H and ADC, are generated from these clocks by clock generators. The implementations of the D-flipflop and the clock generators, employed in Fig. 4.48, are illustrated in Fig. 4.49 [Rab 96, Joh 97].

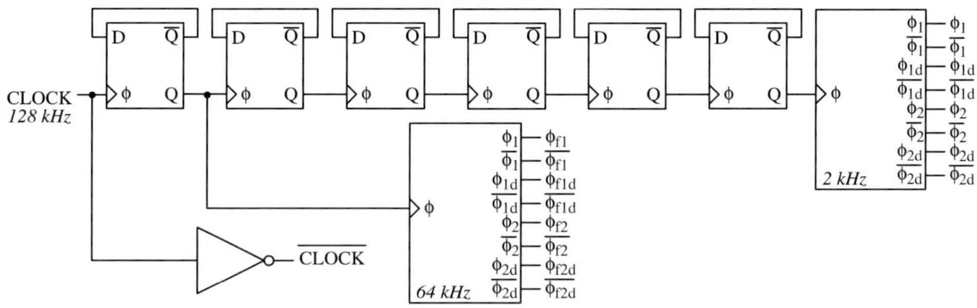

Figure 4.48: *Derivation of the 2-kHz and 64-kHz system clocks, and of the non-overlapping clocks.*

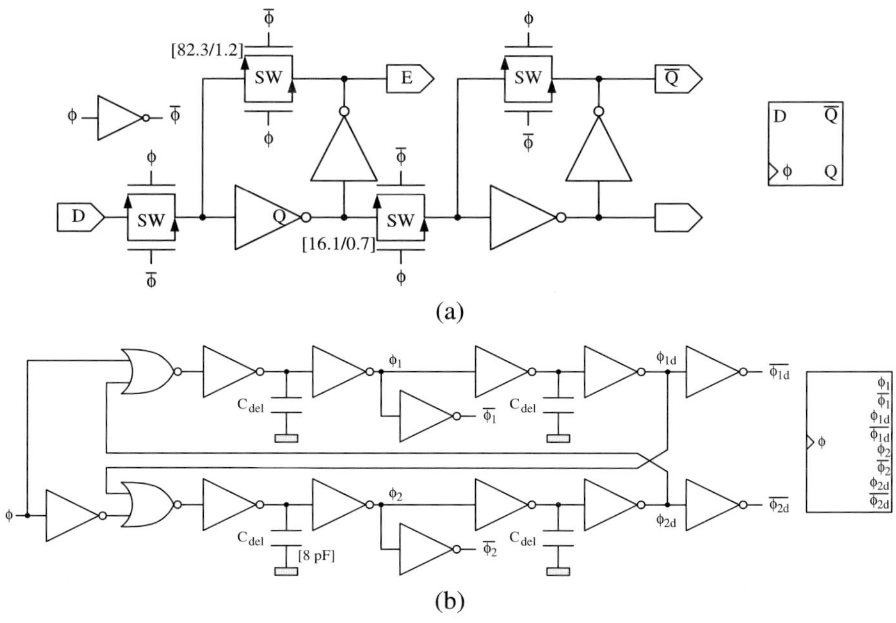

Figure 4.49: *(a) D-flipflop and (b) Non-overlapping clock generator.*

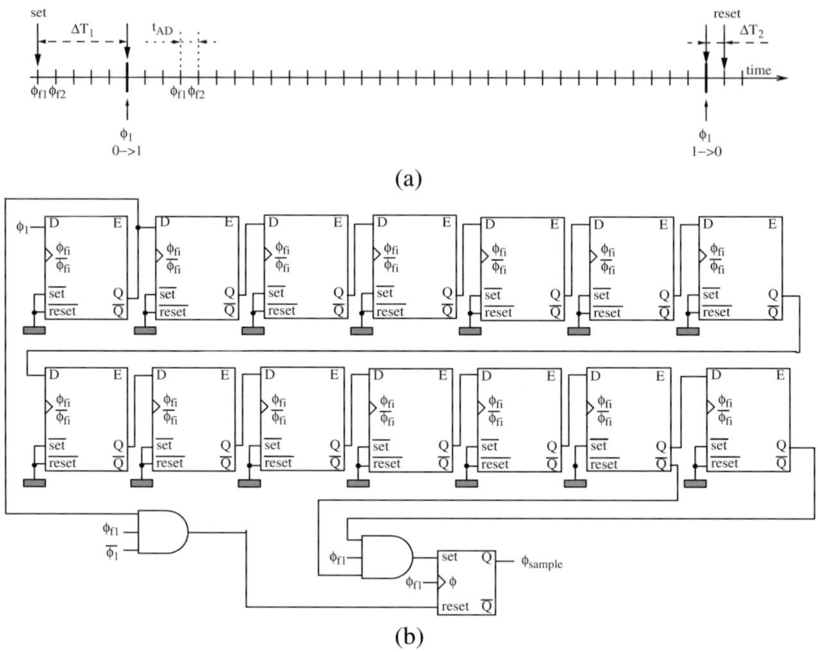

Figure 4.50: *(a) Timing of ϕ_{sample} and (b) Implementation of ϕ_{sample}.*

As already discussed in Section 4.3.4.2 a special clock ϕ_{sample}, derived from ϕ_1, is implemented to clock the current sources I_{REF}, I_{SG} and I_{DAC} (Fig. 4.19). Fig. 4.23 and Fig. 4.50 (a) show the timing of ϕ_{sample} compared with ϕ_1. The implementation of this special clock ϕ_{sample} is illustrated in Fig. 4.50 (b).

4.4 Layout

Fig. 4.51 shows a photograph of the realized 0.7-μm mixed-signal CMOS sensor interface chip, which measures 4.6 mm by 5.2 mm. The most important building blocks, also shown in the schematic overview of Fig. 4.19, are indicated. Special care has been taken to reduce the interference of the digital parts on the analog parts. First of all, the sensitive sensor-signal inputs are implemented on the opposite side of the ADC and the relaxation clock oscillator CLOCK, operating at a switching frequency of 64 kHz and 128 kHz respectively. To limit the disturbance of the clean analog supply voltages of the signal conditioning-and-amplification circuits by the switching activity of the digital circuits and the switches, the well and substrate contacts of both the digital circuits and the switches are connected to separate bond pads. In this way a low impedance return path is ensured for the injected switching currents [Ing 97]. The different

Building block [Duty cycle]	Current consumption (μA)
I_{SOURCE}	46
I_{SG} [19/32]	308
I_{REF} [19/32]	176
I_{DAC} [19/32] + biasing	20 + 48
AMP	103
S/H	29
ADC, V_H and V_L	170 (190)
CLOCK, $\phi_{i(d)}$ and $\phi_{fi\,(d)}$	20
Maximum current [19/32]	920
Minimum current [13/32]	416
Mean current	715
Mean current/channel @ 3.1 V	39.7 (40.8)

Table 4.16: Total current consumption of the different building blocks including biasing.

power supplies of the various building blocks (analog, switches, and digital) are decoupled separately. The decoupling capacitances can be seen in the contour of the chip, and spread over the chip. The C07MA technology consists of a heavily doped bulk and an epi-layer on top in which the circuits are processed. Due to the low-resistive bulk of this technology, digital guard-rings have to be placed very close to the switching noise injectors. Otherwise, they have no effect, since the majority of the injected currents can reach the low ohmic bulk and spread over the entire chip. For a heavily-doped-bulk process, the best result to avoid substrate-coupling induced errors is obtained by mounting the die with a conductive epoxy glue to the lead frame and by using several bond wires to connect the lead frame to the external ground [Ing 00]. Experimental verification has shown that this technique is very efficient. In the final implementation where the chip is placed on a ceramic Al_2O_3 carrier, a metallic plane, connected to the ground, with the size of the chip is foreseen, to which the chip is conductively glued. Moreover, the analog and digital parts are separated by 'grounded' wells. It may be expected that this measure has little effect regarding substrate noise coupling due to the low-resistive bulk, but on the other hand, these 'grounded' wells are useful to shield the signals from the substrate.

4.5 Experimental results

4.5.1 Current consumption

Table 4.16 shows the current consumption of the different building blocks. To reduce the total mean current consumption, the current sources I_{REF}, I_{SG} and I_{DAC} of the multi-gauge nulling block are turned off during most of the amplifier's reset phase ϕ_2. This is performed by the special clock ϕ_{sample} which switches off these currents during 13/32 of the total time. In this

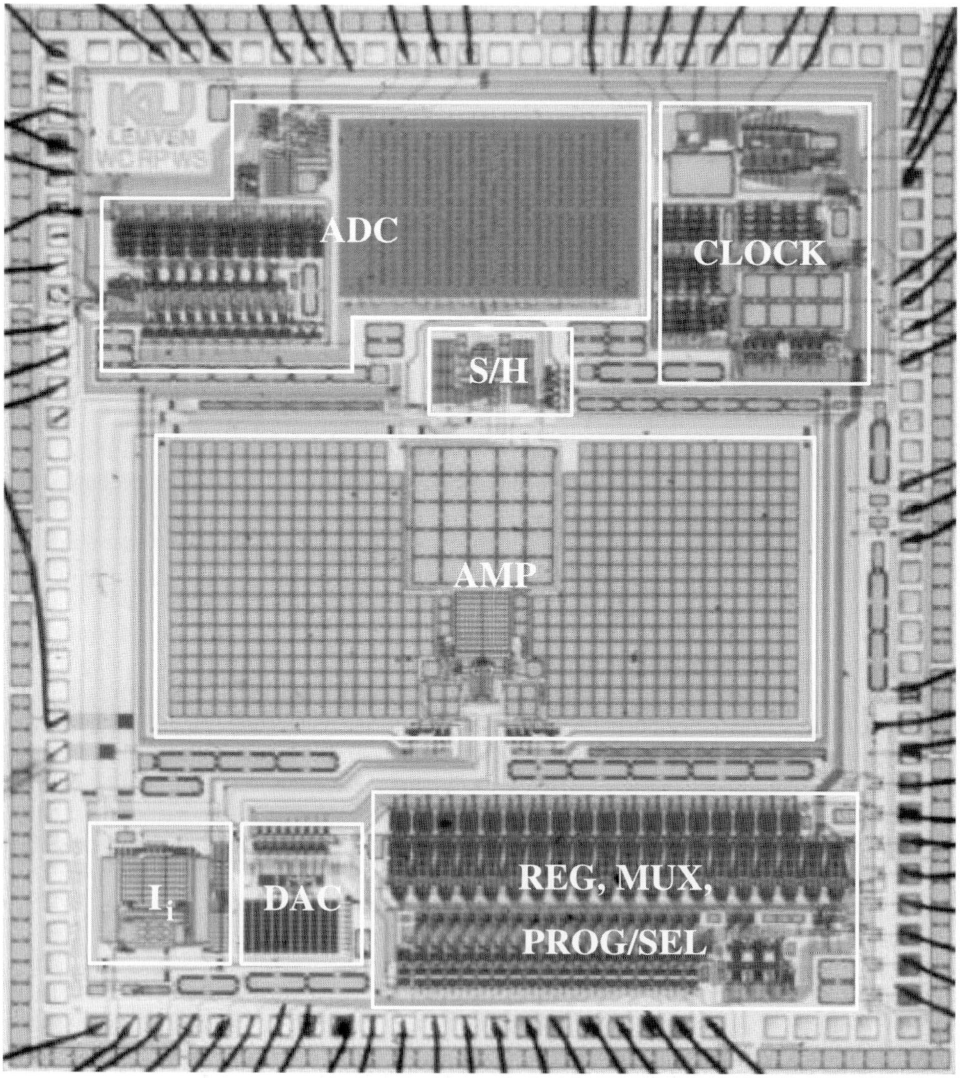

Figure 4.51: *Chip photograph with indication of the most important building blocks.*

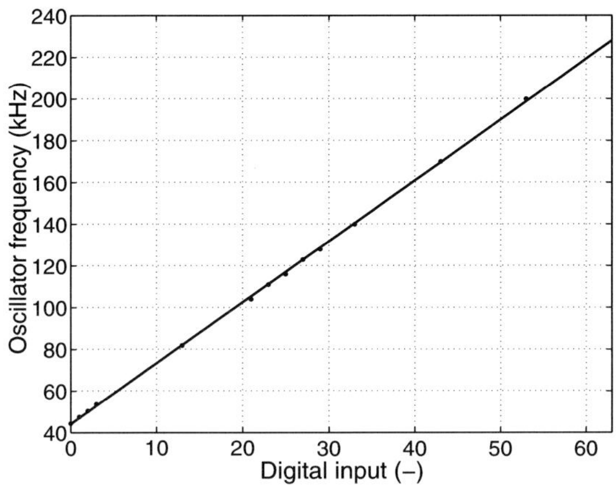

Figure 4.52: *Oscillator frequency measurement for different input codes and least-squares fit.*

way, the resulting mean current consumption of the building blocks is reduced to 39.7 μA (123 μW @ 3.1V) per strain-gauge channel. Note that this current consumption is only valid for the sensor interface chip. The intelligent-datalogger IC, described in the next chapter, uses a slightly different ADC of which the current consumption is given between brackets in the second column of Table 4.16, so that its sensor-interface mean-current-consumption/channel is 1.1 μA higher. The reason for this will become clear further. The Figure Of Merit (FOM) of the final ADC including the biasing, as defined in [Gul 01], equals 0.89 THz/W. The measured mean current consumption of the total chip, including the digital circuitry, is less than 40 μA per strain-gauge channel, which is lower than reported in similar work [Ber 88, Fol 90, Cap 96, Beg 97]. Moreover, the presented system has the additional capability to compensate the different strain gauges and also has offset-cancellation of the different building blocks.

4.5.2 Clock

The performance of the relaxation clock oscillator CLOCK is now briefly discussed. The dots in Fig. 4.52 show the measured frequencies of the clock oscillator for different digital input words. The slope of the least-squares fit, also illustrated in Fig. 4.52, equals 2.92 kHz/bit.

4.5.3 ADC performance

During the testing of the ADC a problem showed up. Digital output codes are missing at the multiples of 32: 4 codes are missing at 256, 2 codes at 128 and 384, and 1 code at the other multiples of 32. Closer investigation reveals that the problem is systematic, since a regular pattern is found in the missing codes. The problem is chip-independent and frequency-independent, but dependent on the step taken in the successive approximation. Several causes have been

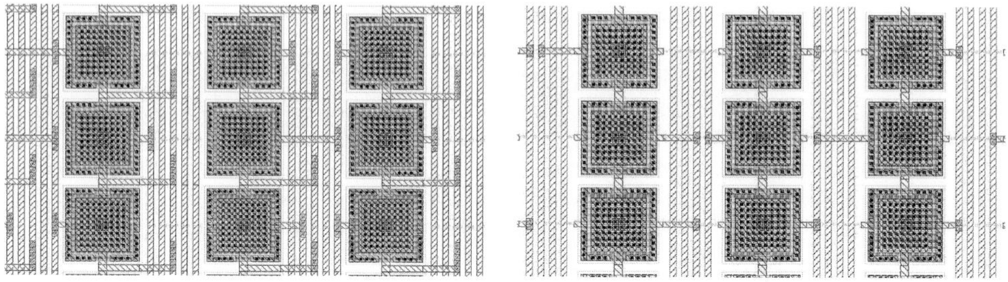

(a) (b)

Figure 4.53: *Unit capacitor array of (a) the sensor interface chip and (b) the datalogger IC.*

investigated: parasitic capacitances [Sak 83], non-linearity of the capacitors, voltage settling errors and dielectric relaxation [Fat 90], but simulations indicate that none of these is the true cause of the problem.

To solve the problem two measures have been taken:

- The layout of the common-centroid unit-capacitor-array has been slightly changed. The difference between the capacitor array of the sensor interface chip and the datalogger IC (cf next chapter) is shown in Fig. 4.53. The distance between the connections to the unit capacitors, routed in between the unit capacitors, and the unit capacitors themselves has been increased in the datalogger IC. Although the unit capacitors have identical surroundings, via's are used to connect the metal-1 unit-capacitor terminals to the metal-2 connections. When these via's are placed too close to a unit capacitor the capacitance value of that capacitor is influenced by these via's. By increasing the distance between the connections /via's and the unit capacitors, this effect is avoided. To make this possible the top plates of the capacitances form together the connection to the comparator input in the new layout.

- The currents generating the reference voltages V_H and V_L are also increased, resulting in lower resistance values of the reference resistors. In this way an extra safety margin is included for settling (cf Section 4.3.6.5), and the possibility of a voltage-settling-error induced problem is fully excluded.

By taking these measures the problem of the ADC is solved, which has been proven by the datalogger-IC measurements, presented in the next chapter, where missing codes do not occur anymore. Another solution to solve the problem is the addition of a digital correction algorithm, explained in Appendix B. This algorithm has been employed for the measurements of the sensor interface chip. An adjustable version of this algorithm also has been implemented in the datalogger IC. Since the ADC problem has been solved by the measures taken, this correction algorithm is redundant though in the final datalogger IC.

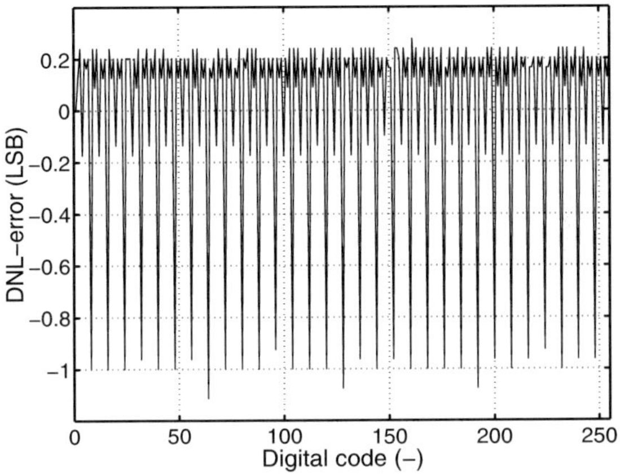

Figure 4.54: *DNL-error DAC.*

4.5.4 DAC performance

Fig. 4.54 depicts the measured DNL-error of the implemented DAC. The systematic errors resulting from the non-perfect current divisions(cf Section 4.3.2.2) are higher than the simulated ones (cf Fig. 4.14), resulting in DNL-errors smaller than -1 LSB for certain digital codes. This however does not impose a problem to achieve accurate nulling, since the DNL-error is smaller than +0.58 LSB over the total range, so that the DAC-accuracy requirement is satisfied. The introduced non-monotonicity for certain codes can cause a possible error in the digital successive approximation algorithm, which is applied in the full datalogger to achieve automatic offset compensation, as described in the next chapter, but this is easily solved by applying some extra 'fine' nulling steps after the 'coarse' nulling has been carried out by successive approximation. This will be discussed in detail in the next chapter (cf Section 5.6.1).

4.5.5 Static measurements

To perform static measurements the measurement setup, shown in Fig. 4.55 (a), has been developed. This pneumatically controlled test setup is able to impose axial forces and bending moments to an abutment under test. The abutment is equipped with strain gauges and fixated to a steel disc with a M2 screw (cf Fig. 4.56 (a)). The steel disc is loaded by the test setup and the applied force is measured by a load cell, and displayed on a LED display [Red] with a resolution of 1 N. The digital output of the sensor interface chip for an increasing bending moment, i.e. an increasing force applied at 1 cm from the center of the abutment, is shown in Fig. 4.55 (b). The slope of the least-squares fit of the measured data is 0.78 $\frac{\text{N}\cdot\text{cm}}{\text{bit}}$ and the standard deviation σ of the measurement error is 1.44 N.cm, which is higher than the required measurement accuracy of 0.79 N.cm (cf Eq. (3.12)). Note that this error results from the inaccuracy of the sensor interface chip *and* the inaccuracy of the measurement setup. The non-Gaussian distribution of

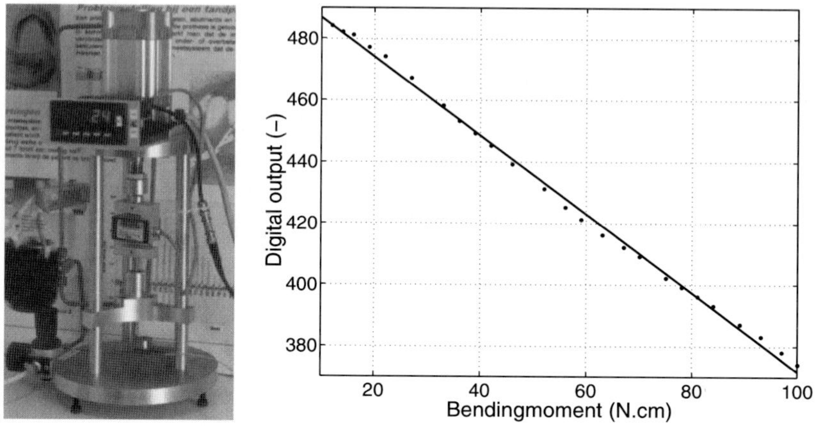

(a) (b)

Figure 4.55: *(a) Static measurement setup and (b) bending-moment measurement.*

the measurement errors indicates that the error is systematic, resulting from the measurement system, which is confirmed in the next chapter, where an alternative measurement setup is used to carry out the static measurements, showing that the static-measurement-accuracy requirement is satisfied with an ample margin.

4.5.6 Dynamic measurements

To carry out dynamic measurements the test setup, shown in Fig. 4.56 (a), has been developed. It consists of a voltage-controlled piezoelectric actuator [Phy b], composed of PZT ceramic stacks, which is driven by a position servo controller [Phy a]. The actuator is able to apply (sinusoidal) displacements to a steel disc that is fixated with a M2 screw to an abutment under test. It is capable to impose displacements with a maximum of 60 μm and with a maximum force of 1000 N. The maximum frequency of the displacements is limited due to the non-fixed connection of the piezo actuator and the steel disc, which introduces non-linearities at higher frequencies. Non-linearities are also introduced by the weak connection of the M2 screw, and the bending of the steel disc itself. In order to avoid these and because of the maximum displacement/pushing force is restricted, the maximum bending moment that can be imposed to the abutment is limited too. In spite of these drawbacks, this test setup permits to perform measurements with real abutments. In this way measurements are carried out which are very similar to the actual measurements of the loads on the prosthesis abutments. In the next chapter measurements performed with an alternative measurement setup with a better linearity will also be presented.

Because of the limited bending moment that can be applied by the test setup, the total ADC range is split up in 19 overlapping intervals and a dynamic measurement is carried out for each of these intervals. The different intervals are selected by changing the digital input of the DAC. For each of the intervals the Power Spectral Density (PSD) of the digital output for a 2-Hz sinusoidal

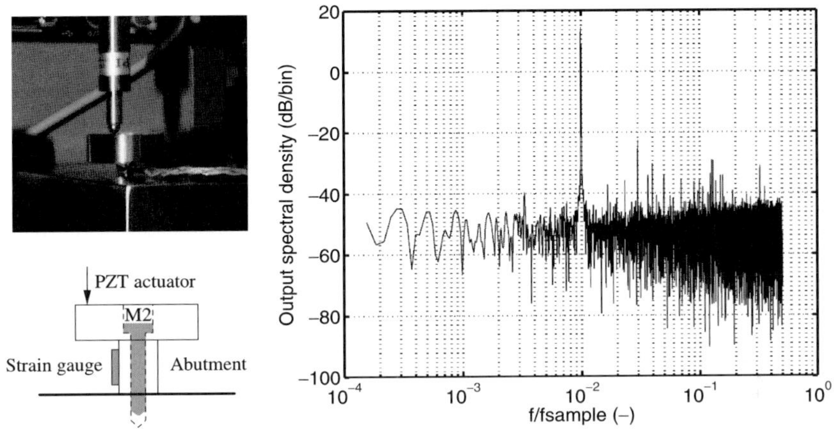

(a) (b)

Figure 4.56: *(a) Dynamic measurement setup and (b) sinusoidal-strain measurement PSD.*

input strain with a peak-to-peak amplitude of 405 μstrain has been derived and the Signal-to-Noise-and-Distortion Ratio (SNDR) has been determined to study the dynamic performance. This yields a mean standard deviation σ of the error of the system (in combination with the mechanical test setup) over all the intervals of 6.2 μstrain. The maximum measured σ of the error is 6.5 μstrain and the minimum equals 5.9 μstrain [Joh 97].

As an example of a dynamic measurement, Fig. 4.56 (b) shows the PSD of the measured output data for a sinusoidal strain with a peak-to-peak amplitude of 336 μstrain and a frequency of 20 Hz. The sample frequency f_{sample} in Fig. 4.56 (b) is equal to 2 kHz. From this PSD the SNDR can be derived, which equals 25.5 dB, corresponding to a standard deviation of the error equal to 6.3 μstrain.

4.6 Conclusion

In this chapter the design of the datalogger's multi-gauge offset-compensated sensor interface, integrated on a separate chip, has been presented. To interface with the strain gauges a current-driven Wheatstone configuration is applied, yielding a lower power consumption in comparison with a voltage-driven one. To cope with the offsets in the different channels introduced by pre-strains and the tolerance on the strain gauges' nominal resistance, the configuration is extended with a digitally-programmable compensation current, implemented by means of a current-steering DAC. The required digital compensation words for every channel are stored in an on-chip nulling memory. Multiplexers are used to switch between the 18 strain-gauge channels and a digital interface is included to program the compensation words in the nulling memory, and to apply the proper compensation word to the DAC when a particular strain-gauge channel is selected.

The complete sensor interface chip includes:

- a *reference current source*. This is a self-biased thermal-voltage-referenced reference current source. The currents of the current-driven Wheatstone configuration are all derived from this mother reference current source. Its dependence on mismatch, supply voltage and temperature has been discussed. In addition to the offsets introduced due to pre-strains and the strain gauges' resistance tolerance, the DAC must also compensate for the offsets, caused by mismatch in the current mirrors, which are used to derive the different currents from the reference current source. Therefore, a formula to calculate the accuracy of a current mirror, depending on the number of transistors in both branches and the input current accuracy, has been derived. This formula has been proven by Monte Carlo simulations.

- a *DAC*, which is an 8-bit binary-weighted current-steering DAC with a maximum DNL-error specification of +0.58 LSB. The LSB bits are realized with series-connected unit current sources, providing a reduction in silicon area of the unit current source array because of the routing of the connections to the unit current sources. The series-connected transistors give rise to non-perfect LSB currents, but it has been proven though that they do not jeopardize the DNL-requirement.

- a *PROG/SEL-block*. This block is used to program the compensation words for the different strain-gauge channels in the on-chip nulling memory. It has two modes: a measurement mode, where it applies the compensation word belonging to the measured strain gauge to the DAC, and a programmation mode, where it is possible to program the compensation words of the different channels.

- a *SC amplifier and S/H*. These are both equipped with offset and 1/f-noise compensation based on the CDS technique. The effect of a finite OTA gain on their performance, their settling behavior and their equivalent input-referred noise are investigated. The noise in SC circuits results from two different propagation methods: direct broadband noise and (aliased) sample-and-hold noise. The input-referred rms noise voltage of the multi-gauge nulling block including the amplifier and S/H equals approximately 8.84 μV. The effects of clock feedthrough and charge injection are reduced by advanced clocks and by making the capacitances as well as the switches seen at the positive and negative OTA input of these building blocks the same. The distortion, CMRR and PSRR related to the SC amplifier, which has a gain of 70, also have been discussed.

- an *ADC*, which is a 9-bit successive approximation ADC with a charge-redistribution DAC with a maximum INL-specification of 0.1 LSB. The preamplifier of this ADC is also offset-cancelled based on the CDS technique. To reduce the power consumption the conversion times of the 3 MSB bits are extended in comparison with the other bits, easing the settling specification for these MSB bits. The building blocks of the ADC have been discussed in detail.

- an *oscillator*, which is a 128-kHz relaxation oscillator. To cope with technology variations the oscillator frequency is programmable by a 6-bit digital word. Two clock dividers (/2 and /64) and two non-overlapping clock generators are implemented to derive the clocks required for the other building blocks. Moreover, a dedicated block is implemented to derive ϕ_{sample}, which is used to switch off the currents of the Wheatstone configuration during most of the reset phase of the SC amplifier, resulting in a reduced power consumption.

The sensor interface chip has been realized in a 0.7-μm CMOS technology and measures 4.6 mm by 5.2 mm. To conclude the measurements of the realized chip have been presented. During the testing of the ADC a problem showed up, which has been solved in the complete datalogger IC, presented in the next chapter, by adaptation of the layout of the charge redistribution DAC of the ADC. The measurements have shown a mean current consumption per strain-gauge channel limited to 40 μA and a dynamic accuracy better than 10 μstrain.

Chapter 5

Intelligent-datalogger IC with programmable data processing

5.1 Introduction

In this chapter the design of the single-chip intelligent-datalogger IC, which has been introduced in Chapter 3, is presented. This chip includes a sensor interface, which has been elaborated in the previous chapter, a digital part and a wireless transceiver. In order to restrict potential errors as much as possible, the sensor interface chip has been tested together with the digital part and the transceiver by means of an FPGA before the actual realization of the datalogger IC.

The principle of operation of the datalogger is described followed by an overview of the building blocks of the digital part and their function. The implemented bi-directional transceiver allows to reconfigure the device and to retrieve collected data and status bytes. The transceiver operation is explained and the list of commands that can be issued to the datalogger is given. The digital part contains automatic offset-compensation towards a user-definable output value for a selectable strain-gauge channel. This is accomplished by means of successive approximation. To reduce the required data storage capacity on board of the datalogger, also a data processing unit with selectable algorithms and programmable parameters is included in the digital part. The communication protocol between the datalogger and the external transceiver is discussed next. An overview of the meaning of the different status bytes, which allow to identify the actual status of the device, is given. After that the layout of the datalogger IC is illustrated and the measurement results are discussed. To conclude the result of a first concept study of the datalogger's packaging is presented.

5.2 Principle of operation

5.2.1 System overview

In Fig. 5.1 (a) the complete bi-directional telemetry system is illustrated again. An overview of the datalogger, incorporated in the dental prosthesis, is shown in Fig. 5.1 (b). It consists of 4

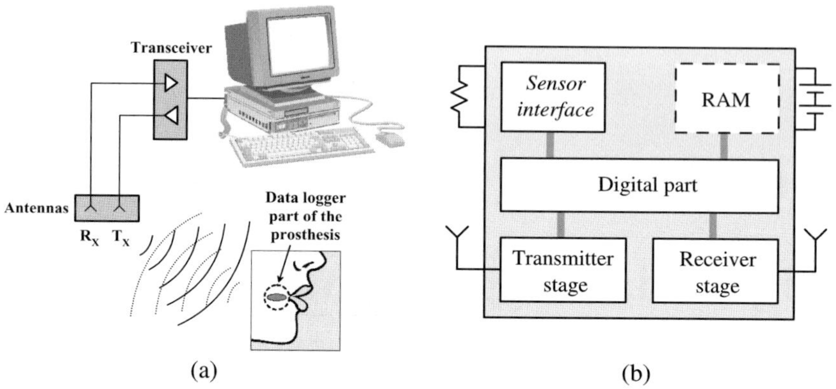

Figure 5.1: *Overview of (a) the bi-directional telemetry system and (b) the datalogger.*

major parts: the sensor interface described in the previous chapter, a digital part, and a wireless transceiver consisting of a receiver and transmitter stage, integrated on a single chip, and an external 2-Mbit SRAM. The sensor interface measures up to 18 strain-gauge sensor signals. The digital part processes the measured digital data and supervises the in- and outgoing communication between the datalogger and the outside world. Furthermore, it takes care of the storage of processed data in the SRAM and the retrieval of stored data from the SRAM, and it controls the operation mode/configuration of the datalogger. It is also capable to perform autonomously the offset-compensation of the different strain-gauge channels as described further. In other words, the digital part adds intelligence to the datalogger. The wireless transceiver is included to adjust the configuration of the device *in situ* and to collect the status bytes, revealing the actual configuration of the datalogger, or the processed data stored in the SRAM wirelessly. In the following sections the functionality and the implementation of the different building blocks are described in detail.

5.2.2 Operation modes

The operation of the datalogger is based on the principle of a transponder [Wou 95]. It is able to pick up a nearby 132-kHz activation field, transmitted by the transmitter T_X antenna (Fig. 5.1 (a)), and respond to it. The transmitter T_X antenna is driven by an external transceiver, controlled by a PC on which runs a dedicated PASCAL program. The remote activation of the datalogger results in an information transfer between the datalogger and the external PC. The receiver R_X and transmitter T_X antennas of the external RF unit are LC-circuits tuned respectively to 66 kHz and 132 kHz. By amplitude modulation of the 132-kHz carrier, transmitted by the RF unit, the datalogger can be (re-)programmed. On the other hand data are transmitted from the datalogger to the receiver antenna R_X by modulation of a 66-kHz carrier, which is derived on board from an incoming non-modulated 132-kHz field, generated by the transmitter antenna T_X. More details about the communication protocol are given in Section 5.4.

The datalogger has two major modes of operation:

- The first mode is the **monitoring mode**. In this mode the sensor interface measures the sensor signals of the different strain-gauge channels and the digital part processes the measured data according to a programmable data processing algorithm (cf Section 5.6.2). The processed data are stored in the SRAM and, as long as the datalogger is not programmed to stop this monitoring by an external activation/programming field, the datalogger continues to collect and process data from the different strain-gauge channels autonomously. As long as the datalogger is in this mode, the bit *monitoringmodeon* (cf Table 5.4), part of the status bytes, is set (high).
- The second mode is the **telemetry mode**. In this mode it is possible to (re-)configure the datalogger and to collect stored data or status bytes from the datalogger. A (re-)configuration of the datalogger is carried out e.g. before the actual measurements start to optimize the data processing algorithm for an individual patient and to activate the automatic offset-compensation, and e.g. at consultation during the measurement period of 2 days. In order to enter the telemetry mode an external activation/programming field must be present. The monitoring must have been stopped before the settings of the datalogger can be reconfigured. This is achieved by programming the datalogger with the correct command (cf Section 5.5). To re-enter the monitoring mode the datalogger must be programmed to start monitoring again.

5.3 Digital part and external SRAM

In this section a global overview of the building blocks of the digital part, and their function, is given. More details of the implementation of the building blocks are given in Section 5.6. The block diagram of the digital part is illustrated in Fig. 5.2. It is implemented as a Finite State Machine (FSM) with VHDL-code and is clocked by the sensor interface 128-kHz clock $\overline{\text{CLOCK}}$ (Fig. 4.48) and the inverse of this clock CLOCK. The latter is used for synchronization purposes as described further. The digital part contains a programmable *data processing* unit including selectable algorithms with adjustable parameters. This unit is implemented to reduce the required data-storage-capacity on board of the datalogger, which is restricted by the available space to incorporate the datalogger in the prosthesis. This unit ensures that only clinical relevant data are stored in the memory. The digital unit also includes a *programming* unit to program the compensation words into the on-chip nulling memory REG via the PROG/SEL-block (cf Section 4.3.3). Moreover, the datalogger is capable to compensate itself for the offsets introduced in the strain-gauge channels. This compensation is carried out automatically by commanding the datalogger wirelessly to compensate towards a user-definable output value for a selectable strain-gauge channel. This is performed by the *nulling* block in combination with the *programming* unit. Successive approximation is employed to determine the required compensation word to be stored into the nulling memory. The automatic nulling is carried out for each channel after placement of the prosthesis before the beginning of the measurements. The *sampling* unit controls the 5-bit channel-select decoder included in the PROG/SEL-block of the sensor interface.

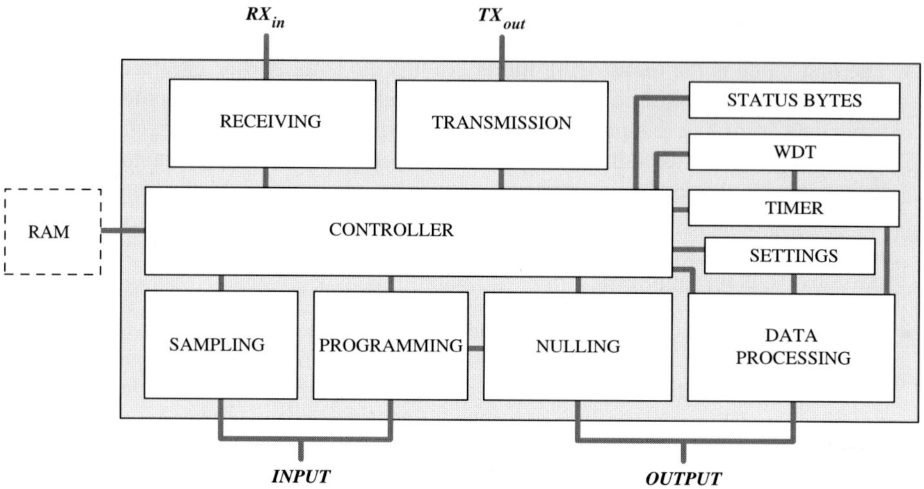

Figure 5.2: *Building blocks of the digital part.*

It ensures that the different strain-gauge channels are measured in the correct sequence. The *sampling* unit also controls the storage of processed data in the SRAM, which is dependent on the number of selected strain-gauge channels for the measurement. Note that there is a delay between the selection of a strain-gauge channel and the availability of the measured data of that channel at the sensor interface output. The *sampling* unit takes this into account when the output data are stored in the SRAM. The *transmission* unit supervises the transmission of stored data in selectable data packages of 256 bytes. To achieve a correct communication, the data bytes are Manchester-encoded (cf Section 5.5) and an extra 3-bit header per byte is added as well as an extra parity byte per data page. The datalogger is programmed with Manchester-encoded 15-bit commands, consisting of a command code of 5 bits and 10 data bits, with an extra 4-bit header and 2 extra parity bits. The *receiving* unit of the datalogger takes care of the reception and validation of these commands and appropriate actions are taken by the *controller* if a correct command is received. The controller orchestrates and supervises the total system. After programming, the actual status of the datalogger can be verified by calling the device *status bytes*. The *controller* also ensures that normal operation is reassumed after a possible lock during communication by means of programmable Watch Dog Timers (*WDT*), as explained further.

The external memory is a 2-Mbit SRAM [BSI], consisting of 262144 locations of 8 bits. It has a STSOP-32 housing, which measures 13.4 x 8 x 1.2 mm^3. Its current consumption is less than 200 μA if it operates continuously at 128 kHz. However, because the memory is put in its power-down mode during most of the measurement period, consuming only 0.15 μA, the mean current consumption of the SRAM over the total measurement period of 2 days is approximately equal to 0.15 μA.

5.4 Transceiver

A simplified overview of the datalogger's integrated transceiver is shown in Fig. 5.3 (a). The external components, placed outside the dotted box, constitute an external rectifier RECT, and a receiver and transmitter tuned LC-tank (RX and TX). The capacitors are SMD components, and the damping resistor R1 of the receiver LC-tank is screen-printed on the Al_2O_3 hybrid carrying the datalogger IC and the other components. The Schottky diode [Wou 95] measures 381x381x150 μm^3 and its bottom (n) is conductively glued to a metallic plane, foreseen on the hybrid, while its top (p) is connected by a bond wire. The cylindrical receiver and transmitter coils L1 and L2 contain a ferrite core and both have an height of 4 mm and a diameter of 4 mm.

The input stage of the transceiver is formed by the tuned LC-tank RX working at 132 kHz. Fig. 5.3 (b) shows the operation of the transceiver during the reception of a command. The external transceiver employs amplitude modulation of a 132-kHz carrier to program the datalogger. The incoming 132-kHz modulated field is picked up by the input stage RX, buffered (BUF) and further rectified over the Schottky diode and the capacitor C5 (RECT). The resulting demodulated signal RX_{in} is processed by the receiver unit of the digital part.

The output stage of the transceiver is formed by the tuned LC-tank TX operating at 66 kHz. During data retrieval of stored data or status bytes from the datalogger, which is shown in Fig. 5.3 (c), a 66-kHz carrier is derived from an incoming non-modulated 132-kHz field. This 132-kHz field is buffered, resulting in a square-wave 132-kHz clock at CL_{TRX}, which is used for both the derivation of the 66-kHz carrier and the timing of the data transmission. The outgoing data, presented at the TX_{out} pin, are modulated by using either phase modulation of the 66-kHz carrier, carried out by EXOR-ing the data with the carrier, or by amplitude modulation of the 66-kHz carrier, carried out with a NAND as described further. The modulated signal is sent to the external transceiver by means of a buffer which drives the transmission LC-tank TX. Note that this buffer has been custom-designed, such that no current flows back from the LC-tank TX into the non-rechargeable battery of the datalogger via this buffer, which would eventually result in malfunctioning of the datalogger.

The transceiver is able to transmit data to the external transceiver over a distance of 30 cm at a maximum data rate of 4 kbytes/s with a mean power consumption of 2.3 mW. A commercially available, external transceiver [Avo] has been modified to communicate with the transponder-type datalogger. The employed Avonwood transceiver-technology has two different internal coils for both reception (RX) and transmission (TX). A drawback of the employed technology is the extra space required for the two coils in comparison with one-coil technologies. On the other hand, no on-chip oscillator is required to generate the 66-kHz transmission carrier. Moreover, the 132-kHz field, from which the 66-kHz carrier is derived, is externally controlled by a crystal oscillator, so that an accurate 66-kHz transmission carrier related to the outside network is obtained, resulting in an enhanced communication.

A more detailed overview of the integrated transceiver is shown in Fig. 5.4. The presence of an external 132-kHz activation field is signaled to the digital part by the *wakeup* pin. When the device is activated by a 132-kHz activation field, the rectified input signal RX_{in} becomes high, resulting in the setting of *wakeup*. Note the introduction of edge-triggered D-flipflops (DFF) in Fig. 5.4 (a) which are employed for synchronization of the (asynchronous) incoming signal

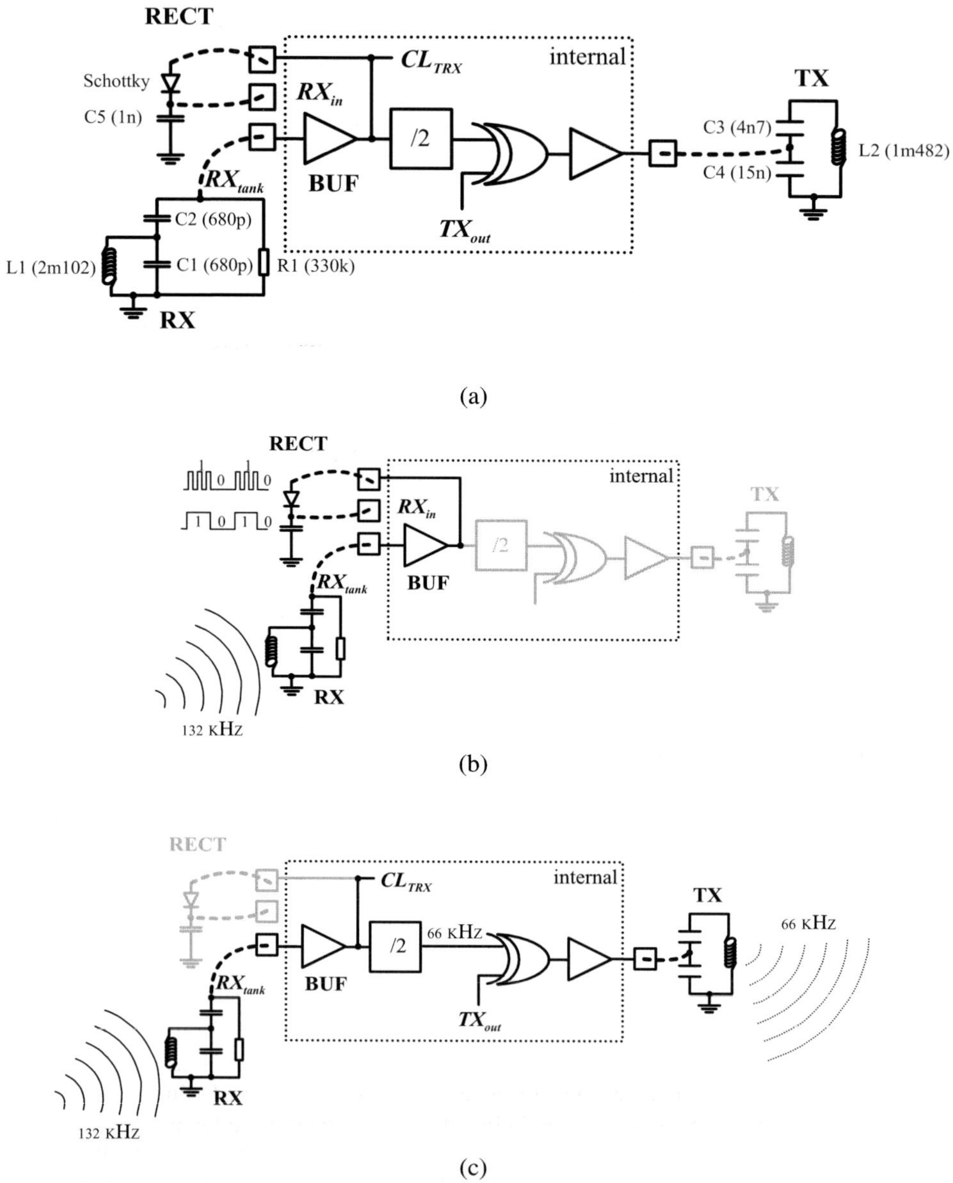

Figure 5.3: *(a) Transceiver (b) during command reception and (c) data transmission.*

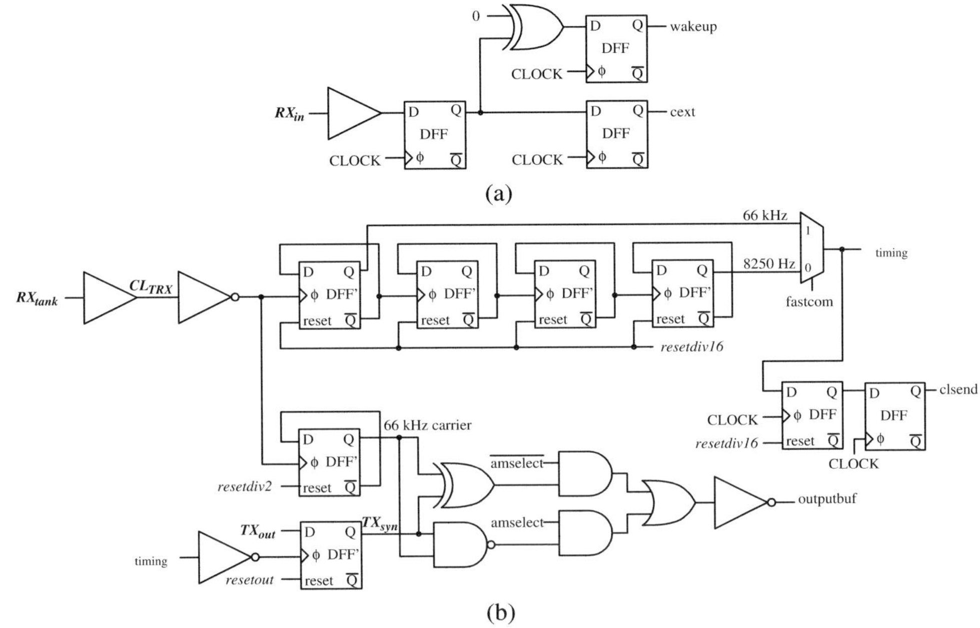

Figure 5.4: *Circuit implementation of the internal transceiver.*

RX_{in}. The synchronizers are clocked by CLOCK, while the digital part of the datalogger is clocked by the inverse of this clock $\overline{\text{CLOCK}}$. This means that RX_{in} is sampled at the rising edges of CLOCK and transferred to the digital part at the rising edges of $\overline{\text{CLOCK}}$ one and a half clock period later. In this way potential errors, related to the flipflops' meta-stability [Rab 96], have time to resolve, confining possible synchronization errors. The *cext* pin in Fig. 5.4 (a) is the synchronized version of the incoming signal RX_{in}, which is further employed by the receiver unit of the datalogger.

During the transmission of data from the datalogger to the external transceiver, a 132-kHz non-modulated field is employed for the derivation of the transmission carrier as well as for the timing of the data communication. The incoming RF signal, presented at the RX_{tank} pin, is buffered, resulting in a square-wave 132-kHz clock at CL_{TRX}, which is divided by a master-slave D-flipflop (DFF') divider to produce the 66-kHz transmission carrier. Moreover, to help the timing of the transmission unit, a 8250-Hz clock signal and a 66-kHz clock signal are derived from the 132-kHz field by means of 4 edge-triggered master-slave D-flipflop dividers. The selection between the two clocks for the timing of the data communication is carried out by the *fastcom* pin, which controls a multiplexer. The *fastcom* pin can be set/reset by the transmission-mode command (cf Table 5.1 and Table 5.3). The synchronized version *clsend* of the selected clock ϕ_{timing} is employed by the transmission unit, as explained further. $\overline{\phi_{\text{timing}}}$ on the other hand is used to synchronize the outgoing data, presented at TX_{out}, and the 66-kHz transmission carrier. Since the outgoing data only can change at positive edges of $\phi_{timing}/clsend$ (cf Section 5.6.4) and the synchronizer samples at negative edges of this clock, potential problems related to metastability are restricted. The integrated transceiver leaves the choice between Phase Modulation ($\overline{amselect}$) or Amplitude Modulation (*amselect*) for the outgoing data. The latter is implemented by NAND-ing the outgoing data with the carrier. The selection between AM and PM is also carried out by the transmission-mode command (Table 5.1). Note that the different flipflops can be reset by *resetdiv16*, *resetdiv2*, and *resetout* respectively.

As already mentioned above, a commercially available, external transceiver [Avo] has been modified to communicate with the datalogger. This transceiver is normally employed to read out/program RFID tags, which are for example used in the animal husbandry. A dedicated communication protocol for the RFID tags is implemented with custom(-programmed) chips on the transceiver PCB. These chips are removed from the PCB and only the real transceiver part of the PCB is utilized for this application. The latter is connected to the parallel port of the PC on which runs a dedicated PASCAL-program to program/read out the datalogger, and to store and visualize the collected data. A drawback of this approach is that the data transfer rate is limited due to the limited speed of the PC's parallel port. On the other hand, this approach allows fast prototyping of the software to communicate with the datalogger. In order to achieve data collection at full speed, in the future the transceiver PCB must be combined with a microcontroller containing the assembler code of the program developed in PASCAL.

5.5 Instruction set

Table 5.1 gives an overview of all the commands that can be issued to the datalogger. Each command consists of a 5-bit command code and 10 data bits. The latter are used to program the parameter values belonging to a command. More details about the meaning of the data bits are given further. The list of commands in Table 5.1 is divided into 5 subgroups:

- I. Commands related to the general operation of the device. These commands allow to start and stop the monitoring mode, and to reset the settings of the datalogger.

- II. Commands employed to store the digital compensation words into the nulling memory REG 'manually' or to activate the automatic nulling of a selectable strain-gauge channel towards a selectable output value.

- III. Commands used to set up the employed data processing algorithm and to select the number of strain-gauge channels which are monitored during a measurement.

- IV. Commands to set up the communication mode and to activate the transmission of a selectable data page or the status bytes. The latter allow to retrieve the actual status of the datalogger and their number is equal to 21. The collection of stored data on the other hand is carried out in packages of 256 bytes, also called data pages.

- V. Command for the programmation of the WDT intervals.

Code	Command description	Class
10000	Reset datalogger	I
00000	Stop sampling	
01111	Start sampling	
00010	Activate nulling/programming of a selectable strain-gauge channel	II
00001	Program nulling word/programming word	
00110	Program first and last strain-gauge channel	III
00101	Select data processing algorithm	
00111	Program thresholds of data processing algorithms	
00011	Reset timer	
00100	Calibrate timer	
01000	Program sample counter value	
01011	Program transmission mode	IV
01001	Program single bit transmission length	
01110	Activate transmission of a selectable data page with parity byte	
01101	Activate transmission of (21) status bytes with parity byte	
01010	Program WDT intervals	V

Table 5.1: Overview of the datalogger commands.

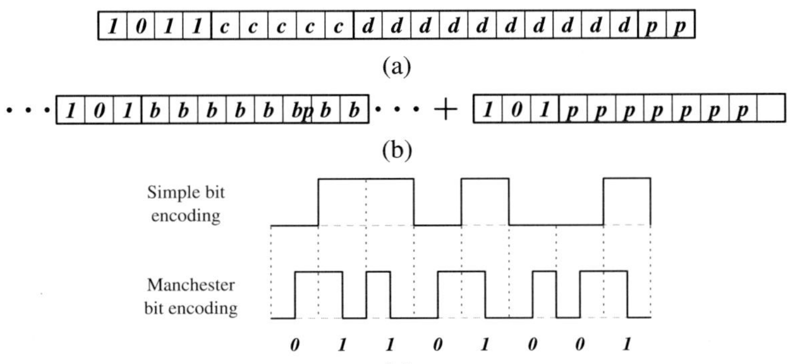

Figure 5.5: *(a) Command format. (b) Data format. (c) Manchester encoding.*

Fig. 5.5 (a) illustrates the format of the commands issued to the datalogger. They consist of a 4-bit header (1011), a command code of 5 bits, 10 data bits, and 2 extra parity bits. The latter correspond with the binary representation of the remainder of the division of the total number of ones in the command code and the data bits by 4. The inclusion of the parity bits allows to check for the correctness of the communication. The format of the data, retrieved from the datalogger, is given in Fig. 5.5 (b). A 3-bit header (101) per transmitted byte is added as well as one extra parity byte at the end of the collection of a data page or the status bytes in order to enhance the correctness of the information exchange. The parity byte corresponds with the binary representation of the remainder of the division of the total number of ones in all the data bytes/status bytes and the 3-bit headers by 256.

The employed encoding scheme is the Manchester-encoding scheme [Wou 95], which is shown in Fig. 5.5 (c). With this scheme a logic 0 is encoded into a half-bit low-to-high transition and a logic 1 in a half-bit high-to-low transition. An advantage of this scheme is that it is self-clocking, since every bit is associated with a half-bit transition. It allows the receiver to be conscious of every bit that is received. This results in an improved information exchange in comparison with a simple bit encoding scheme (cf Fig. 5.5 (c)). When a command is issued to the datalogger, the 132-kHz clock of the external transceiver is also used for timing by the PASCAL program. A half bit corresponds with a programmable period of maximum 1024 cycles of this clock.

5.6 Building blocks of the digital part

5.6.1 Programming and nulling units

In this section the building blocks of the digital part of the datalogger are elaborated further. These blocks have been implemented in VHDL-code and have been tested with a Field-Programmable-Gate-Array (FPGA) in combination with the earlier described sensor interface chip. First the programming unit and the nulling unit are discussed. The programming unit is used to program a selectable register of the sensor-interface nulling memory REG, belonging to a particular strain-gauge channel, with a compensation word *inputDAC*, by means of the PROG/SEL-block. The protocol to program this compensation word into the register of REG has already been discussed in Section 4.3.3. This protocol is implemented in the programming unit, and is initiated when *startprog* is set high (cf Fig. 4.18). On the other hand the nulling unit performs, in combination with the programming unit, automatic nulling of a selectable strain-gauge channel towards an adjustable nulling level *nullingwanted*. It employs a successive approximation algorithm to determine the required compensation word *inputDAC* to be stored in REG. The nulling unit searches for the sensor-interface DAC-input *inputDAC*, which results in a sensor interface output *outputADC* that is larger than the wanted nulling level and that lies within an interval Δ_{nulling}, which corresponds with $[1+\text{DNL}_{\text{max}}]$ LSB_{DAC} where DNL_{max} is the maximum positive DNL-error of the DAC and LSB_{DAC} is the interval between two successive DAC outputs for an ideal DAC (cf Section 4.3.2.1). Δ_{nulling} corresponds with the maximum output step of the DAC between two successive codes.

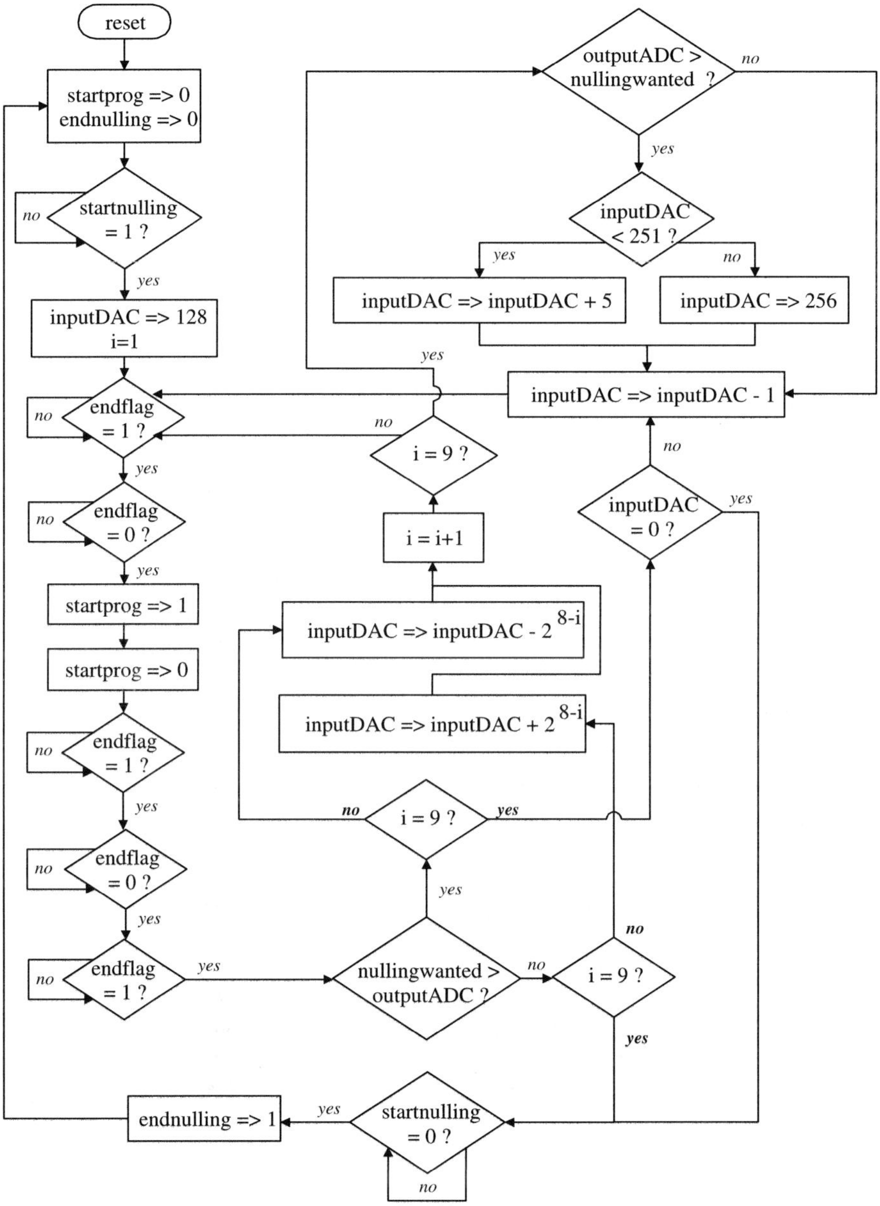

Figure 5.6: *Flowchart of the nulling algorithm.*

The flowchart of the successive-approximation nulling-algorithm is shown in Fig. 5.6. Transitions in this flowchart occur at low-to-high transitions of the 128-kHz clock $\overline{\text{CLOCK}}$. The automatic nulling is initiated by setting *startnulling* high. First, the MSB bit of the compensation word is determined. Therefore, *inputDAC* is set equal to 128 and this value is programmed in the register belonging to the strain-gauge channel for which the nulling is carried out. After a high-to-low transition of *endflag* the programming is initiated by setting *startprog* high. Because of the delay between the change of the DAC input and the resulting change of the ADC output, one conversion-end, signaled by a low-to-high transition of *endflag*, is skipped. At the end of the next conversion, *outputADC* corresponds with the freshly-programmed compensation word *inputDAC*. To determine the next bit of the compensation word, *nullingwanted* is compared with *outputADC* and a new value for *inputDAC* is calculated/programmed depending on the comparison result (cf Section 4.3.6.1). This process is repeated 8 times until all the bits of the compensation word are determined. In each iteration the step applied to the DAC input is divided by 2. Note that an increase of the DAC input results in an decrease of the ADC output, except for certain DAC input codes, which have a DNL-error smaller than -1 LSB_{DAC} (cf Section 4.5.4), resulting in a non-monotonous output of the DAC for these codes.

At the end of the successive approximation, when *i* is equal to 9, *outputADC* lies within an interval corresponding with $\pm\ \frac{1}{2}\Delta_{\text{nulling}}$ from *nullingwanted.* An extra 'fine' nulling step is performed to ensure that *outputADC* is larger than *nullingwanted.* As already mentioned in Section 4.5.4, the DNL-error of the DAC is smaller than -1 LSB_{DAC} for certain digital codes. Due to the resulting non-monotonicity for these codes an error can be introduced in the successive approximation. By applying the extra 'fine' nulling step this potential error is also solved. Depending on the relative size of *outputADC* and *nullingwanted*, *inputDAC* is increased with 4 (if possible) or decreased with one. Next, this value is programmed and the resulting ADC output is compared with *nullingwanted.* If *nullingwanted* still is larger than *outputADC*, *inputDAC* is decreased with one. This process is repeated until the first compensation word is found which gives an ADC output which is larger than *nullingwanted.* The end of the nulling is signalled to the controller by *endnulling*.

The commands with command code **00010** and **00001** (Table 5.1) both are employed for the configuration/initiation of the programming/nulling. The value of the MSB data bit of the commands (cf Fig. 5.5 (a)) determines which action is performed. First the **00010**-command is issued to the datalogger. If the MSB data bit of this command is high, the datalogger uses the 8 LSB data bits as the compensation word *inputDAC* to be programmed into REG. However, if the MSB data bit is low, the other 9 data bits correspond with *nullingwanted.* On the other hand, the **00001**-command is used to activate the programming/nulling. If the MSB data bit of this command is high, programmation of a register is carried out, while in the other case, nulling is performed. In both cases the 5 LSB data bits correspond with the strain-gauge channel *strgnrprog*, for which the programming/nulling needs to be carried out.

5.6.2 Data processing unit

The next building block is the data processing unit. Without data processing onboard of the datalogger a memory capacity of 388.8 Mbytes would be required to store all the 9-bit sensor-

No	Raw	Mean	Thresholds	Time	Max	Duration	MSB data bits
0	X						00000
1		X					00001
2	X		X				00010
3		X	X				00011
4	X		X	X			01010
5		X	X	X			01011
6		X	X	X	X		01110
7			X	X	X	X	11111

Table 5.2: Properties of the implemented data processing algorithms.

interface output data collected during a two-day-period measurement of 18 strain-gauge channels. Because of the limited space available for the datalogger, which is embedded in the dental prosthesis, the size of the onboard memory is restricted. Note that this limitation of the the memory size is beneficial for the power consumption. As already mentioned above, a 2-Mbit memory [BSI] with a STSOP-32 housing, measuring 13.4 x 8 x 1.2 mm^3, has been chosen for the external memory. Due to the limited storage capacity of 2 Mbit data processing onboard of the datalogger is a requisite.

The implemented data processing unit contains 8 selectable algorithms. Moreover, the parameters of these algorithms are programmable. In this way a very flexible data processing unit has been realized, which can be optimized for each individual patient and research domain. Table 5.2 gives an overview of the properties of the 8 selectable algorithms with adjustable parameters (in *italics*):

- *0.* The first algorithm continuously stores the raw data of the different channels without any further processing. This algorithm does not allow memory savings, which results in a maximum data collection time of 116.5 seconds for 18 channels. It is used in the learning cycle to retrieve patient-specific loading behaviour and can be employed to derive the optimal parameters for the other algorithms.

- *1.* The second algorithm is similar to the first one. It stores continuously the average values of the data and uses 2^{nrdata} data points with $1 \leq nrdata \leq 10$ to calculate the averages in each channel. The averages are found by shifting the sum of 2^{nrdata} measured data points by *nrdata*. The data collection time is extended to $2^{nrdata} \cdot 116.5$ seconds, since only the average values are stored now.

- *2.* The third algorithm stores the raw data of the different channels after the output of one of the channels has become smaller than the lower threshold *lowerthreshold* or larger than the upper threshold *upperthreshold*. After trespassing one of the thresholds 2^{nrdata} of raw data points are stored in the memory for each channel.

- *3.* This algorithm is similar to the previous one with the exception that the average values of the data points are stored now after trespassing one of the thresholds. For each channel 2^{nrdata} data points are used for the average calculation.

- *4.* The data processing performed by this algorithm is similar to that of algorithm 2, but now also the time of occurence is stored in the memory. This permits to retrieve the times of occurence when the data are collected from the datalogger, which provides additional information to the dentist. Two timers are integrated in the datalogger: a coarse timer and a fine timer, which are both reset at the beginning of the measurements. The time information stored by this algorithm is the value of the coarse timer. More information about the timers is given further.

- *5.* This algorithm is identical to algorithm 3, but now also time information is stored.

- *6.* If this algorithm is selected, the data processing is carried out in a similar way as by the previous algorithm, but with the addition of the storage of the maximum data set. This data set contains the maximum output that has occured in one of the channels between the trespassing of the upper threshold in one of the channels, triggering the data processing, and the storage of the average values. Each time the output of one of the channels is larger than the previous maximum, the outputs of the other channels are measured and combined with the new maximum into the maximum data set. The final maximum data set is stored in the memory after storage of the average values. Note that the described data processing only can be triggered by the trespassing of the upper threshold. This is different from the previous algorithm, where the data processing is triggered by trespassing either *upperthreshold* or *lowerthreshold*.

- *7.* The last algorithm stores the duration during which the upper threshold is trespassed, the time of occurrence and the maximum data set occured during the trespassing. The duration during which the upper threshold is trespassed is measured with a unit of time equal to the period of ϕ_{sample} multiplied by 18 $\approx$ 9 ms. A 9-bit word is employed to store the duration of the trespassing, which is consistent with the 9-bit format of the processed data provided by the algorithms.

As already mentioned above, a coarse timer and a fine timer are implemented in the datalogger. The coarse timer is used by the data processing unit, while the fine timer is employed for the WDTs as explained further. Fig. 5.7 shows the flowchart of the timers. The basic clock ϕ_{timer}, from which they are derived, is realized by the division of the 128-kHz clock $\overline{\text{CLOCK}}$ by 2^{17} by means of 17 D-flipflop dividers. One period of ϕ_{timer} corresponds with 1024 ms. The fine timer *finetimer* counts the number of periods of ϕ_{timer}, while the coarse timer counts the number of overflows, which occur when *finetimer* becomes equal to *finetimerend*. *finetimerend* is an 8-bit word which can be programmed by the **00100**-command, and is used to calibrate the coarse timer. One unit of time of the coarse timer is equal to 1024 ms multiplied by *finetimerend* and the maximum time between two overloads of *finetimer* equals 256x1024 ms. To reset both the timers the **00011**-command must be issued to the datalogger.

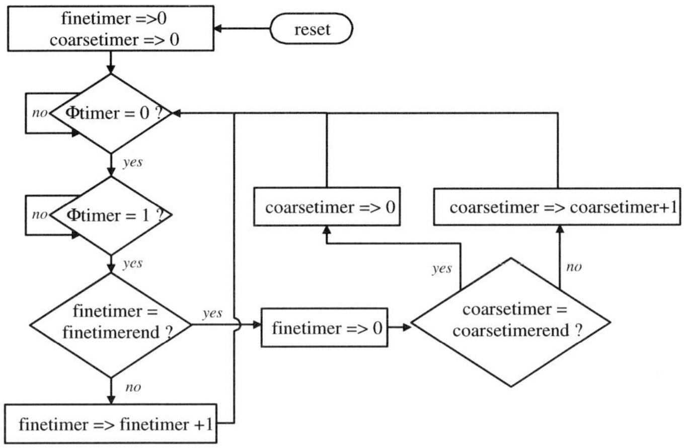

Figure 5.7: *Flowchart of the coarse and fine timer.*

To be consistent with the sensor-interface output-data format and the format of the processed data, *coarsetimer* also consists of 9 bits, which facilitates its storage in the external memory. This means that the maximum time that can be measured without an overload of *coarsetimer* equals 512x256x1024 ms, which is approximately equal to 37.3 hours. From this follows that an overload will occur during a two-day measurement. When the overload occurs, *coarsetimer* is reset. This must be taken into account during the interpretation of the timer data after collection of the data from the datalogger. This does not impose a problem, since in normal conditions the patient eats several times a day, more than likely triggering the data processing.

The selection of the data processing algorithm and the programmation of its parameters are carried out by means of the commands **00101** and **00111**. The 5 MSB data bits of the **00101**-command determine which data processing algorithm is selected. The right column of Table 5.2 illustrates the values of the MSB data bits required to select an algorithm. The 5 LSB data bits on the other hand determine the value of *nrdata*. To program the threshold parameters the **00111**-command must be sent to the datalogger. When the MSB data bit of this command is high, the other 9 data bits correspond with *upperthreshold*, while in the other case, they correspond with *lowerthreshold*.

The number of monitored strain-gauge channels is dependent on the number of implants for a given patient. The selection of the first and the last strain-gauge channel to be measured is done with the **00110**-command. When this command is issued to the datalogger the 5 MSB data bits correspond with the first strain-gauge channel *strgnrfirst*, while the 5 LSB data bits correspond with the last channel *strgnrlast*. These two channels and all the channels in between are selected for monitoring. Note that the data processing algorithms only generate data for these channels. When the processed data are available, this is signalled to the sampling unit by the *dataready* pin.

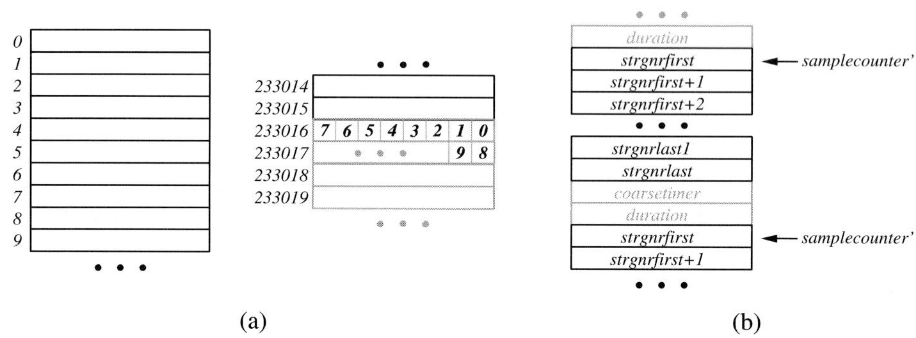

Figure 5.8: *(a) Storage of the 9-bit processed data in the 8-bit memory and (b) storage pattern of the processed data.*

5.6.3 Sampling unit

The sampling unit has two major functions:

• **Channel selection:** The sampling unit controls the 5-bit channel-select decoder included in the PROG/SEL-block of the sensor interface (cf Section 4.3.3). It ensures that the different strain-gauge channels are selected in the correct sequence, corresponding with their channel number. Once the last channel (17) has been reached, the selection cycle starts again from the first channel (0). When the end of a conversion is signalled by a low-to-high transition of *endflag*, a new strain-gauge channel is selected by the sampling unit once *endflag* has been reset by the sensor interface. The 18 different strain-gauge channels are selected at a frequency of ϕ_{sample}, independent of whether they are selected for monitoring or not. In this way the time between two consecutive samples of the same channel is constant, i.e. $18/\phi_{\text{sample}}$, or in other words, the sample frequency in each channel equals 111 Hz. Note that data are only stored for the selected strain-gauge channels, while the selection of the channels by the sampling unit is independent of these. The possibility to put the sensor interface in a power-down mode for the channels, which are not selected for monitoring, has not been foreseen. The datalogger is designed to operate during a two-day measurement period for 18 channels. If a channel is not selected for monitoring the power consumption during the measurement of that channel is reduced, because no strain gauge is present at the sensor interface input, so that I_{SG} and I_{DAC} (Fig. 4.19) can not flow. From this follows that, if the datalogger operates over a two-day measurement period for 18 channels, the operation over the same period is certainly guaranteed for a restricted number of channels.

• **Data-storage supervision:** The sampling unit supervises the storage of processed data, provided by the data processing unit, in the external SRAM. The processed 9-bit data are put in the 8-bit SRAM as illustrated in Fig. 5.8 (a). The LSB bits of the data are stored from memory location 233016 onwards. In this way all the LSBs of the processed data, of which the MSB bits are put before memory location 233016, can be located in last part of the memory. As illustrated in Fig. 5.8 (a) the first LSB of memory location 233016 corresponds with the LSB of the data

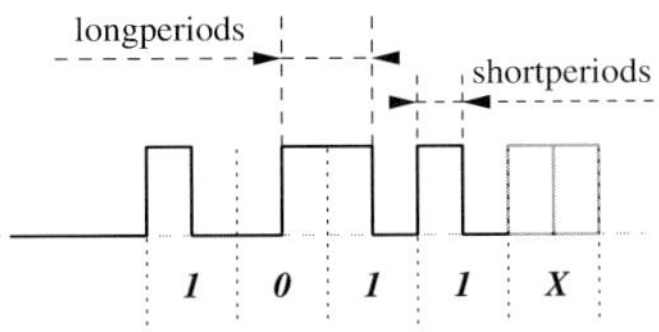

Figure 5.9: *Encoding of the command header and illustration of the decoding timing-parameters.*

put in memory location 0, the second LSB bit with the data put in memory location 1, and so on.

The sampling unit also controls the storage pattern of data in the memory, which is dependent on the number of selected strain-gauge channels and the selected data processing algorithm. Note that there is a delay between the selection of a strain-gauge channel and the availability of the processed data of that channel. The sampling unit takes this into account when these data are stored in the SRAM. Fig. 5.8 (b) illustrates the storage pattern of the processed data. A cluster of data consists of the processed data, starting with data from the first channel *strgnrfirst* and ending with data from the last *strgnrlast*. Note that the boxes, indicated in grey, are optional data, included in a data cluster, dependent on the selected processing algorithm. If one of the 3 last algorithms of Table 5.2 is selected the coarse timer is included, and if the last algorithm is selected the duration of the trespassing is included too. Furthermore, if algorithm 6 is selected, two clusters of data are stored at the end of each data processing: one with the average values and one with the maximum data set. Note that the data storage is triggered by the setting of *dataready*, which is done by the processing unit.

An 18-bit counter *samplecounter*, whose value corresponds with the 18-bit memory-location address to begin the storage of the next data cluster, has been implemented in the sampling unit. Fig. 5.8 (b) illustrates two possible values of *samplecounter*. The counter can be reset or its initial value can be programmed by the user in order to store the processed data from an arbitrarily-chosen memory-location onwards. If *samplecounter* is reset at the beginning of a measurement, its value corresponds with the number of collected data during the measurement. The value of the counter is part of the status bytes. When the status bytes are retrieved from the datalogger, the number of data collected so far can thus be found. To program the initial value of *samplecounter* the **01000**-command is used. If the MSB data bit of this command is high, the 9 other data bits correspond with the 9 MSB bits of *samplecounter*, while in the other case, they correspond with the 9 LSB bits of the counter.

The VHDL code of the sampling unit is given as an example in Appendix C.1, while its flowchart is illustrated in Appendix C.2.

5.6.4 Receiving and transmission units

In this subsection the building blocks, related to the exchange of information between the datalogger and the external transceiver, are discussed. First the receiver unit, of which the flowchart is shown in Fig. 5.10, is elaborated. As explained in Fig. 5.5 (a), a 4-bit header **1011** is added to each command that is issued to the datalogger. This is done to improve the communication, and to derive the timing parameters for the decoding of the Manchester-encoded commands onboard. The input signal of the receiver unit is the rectified-and-synchronized signal *cext* (cf Fig. 5.4). The unit is triggered by *wakeup*, which goes high when the first half of the first bit of the header (i.e. **1**) is transmitted. After *cext* has reset again, the receiver unit commences with the derivation of the two decoding timing-parameters *longperiods* and *shortperiods*, which are expressed in periods of $\overline{\text{CLOCK}}$, i.e. the clock of the receiver unit. As depicted in Fig. 5.9, *longperiods* corresponds with the period between two mid-bit transitions, which occurs if the previous bit is different from the next, i.e. e.g. for **01**. On the other hand *shortperiods* corresponds with the period between the border of a bit, indicated by the dotted lines in Fig. 5.9, and its mid-bit transition. This time interval is present between two transitions if the previous bit is the same as the next, i.e. e.g. for **11**. From these two periods a mean period *meanperiods* is calculated to decipher the incoming command. By the derivation of *meanperiods* from the command header, the datalogger is able to automatically adapt its deciphering in accordance with the length of a half bit of the incoming command signal. In this way no problems are expected when the PASCAL program is replaced by a microcontroller, as explained in Section 5.4, which permits a shorter half-bit length. Note that the WDT, which is explained further, is started after the receiver unit wakes up by the setting of *startlock*.

As illustrated in Fig. 5.10, the receiver unit measures the period *periods*, expressed in periods of $\overline{\text{CLOCK}}$, between two transitions of *cext* and compares this time interval with *meanperiods* to decode the bits of the incoming command. The value of the next bit depends on the value of the previous bit, the value of *periods* and on whether the first transition is a mid-bit transition, denoted by *midi*, or a border transition, denoted by *bori*. If the measured period is larger than *meanperiods* and the first transition is a mid-bit transition, the next bit is different from the previous one and the second transition is again a mid-bit transition, while in the other case the next bit is the same and the second transition is a border transition. If the first transition is a border-transition the period must be smaller than *meanperiods*. Otherwise an error, denoted by *err*, occurred, since the Manchester encoding of the incoming command ensures that a mid-bit transition occurs for every bit (cf Section 5.5). On the other hand, if the period is smaller than *meanperiods*, the second transition is a mid-bit transition. If a mid-bit transition occurs, the value of that bit is put in *command*, i.e. a 17-bit word containing the command code, the data bits and the parity bits. *bitnr* indicates the position of the bit that is being decoded.

At the end of the command reception the external field must be deactivated, reactivated and deactivated again by the external transceiver. This results in the setting of *stoplock*, which ends the WDT operation. The end of the reception is signalled to the checkcommand unit, which is part of the receiver unit, by the *commandreceived* pin. The checkcommand unit calculates the parity bits of the received command and checks whether a valid command is received or not. The parity bits, i.e. the two LSB bits of *command*, must correspond with the binary representation of the remainder of the division by 4 of the total number of ones in the command code and the data bits, i.e. the 15 MSB bits of *command*. If a valid command is received, this is signalled to the controller, which supervises the execution of the received command, dependent on the data bits. If a bad command is received, the *badcommand* bit, which is part of the status bytes (cf Table 5.4), is set. If in this case the status bytes are retrieved from the datalogger after the command reception, the user is informed that an error occured during the last communication.

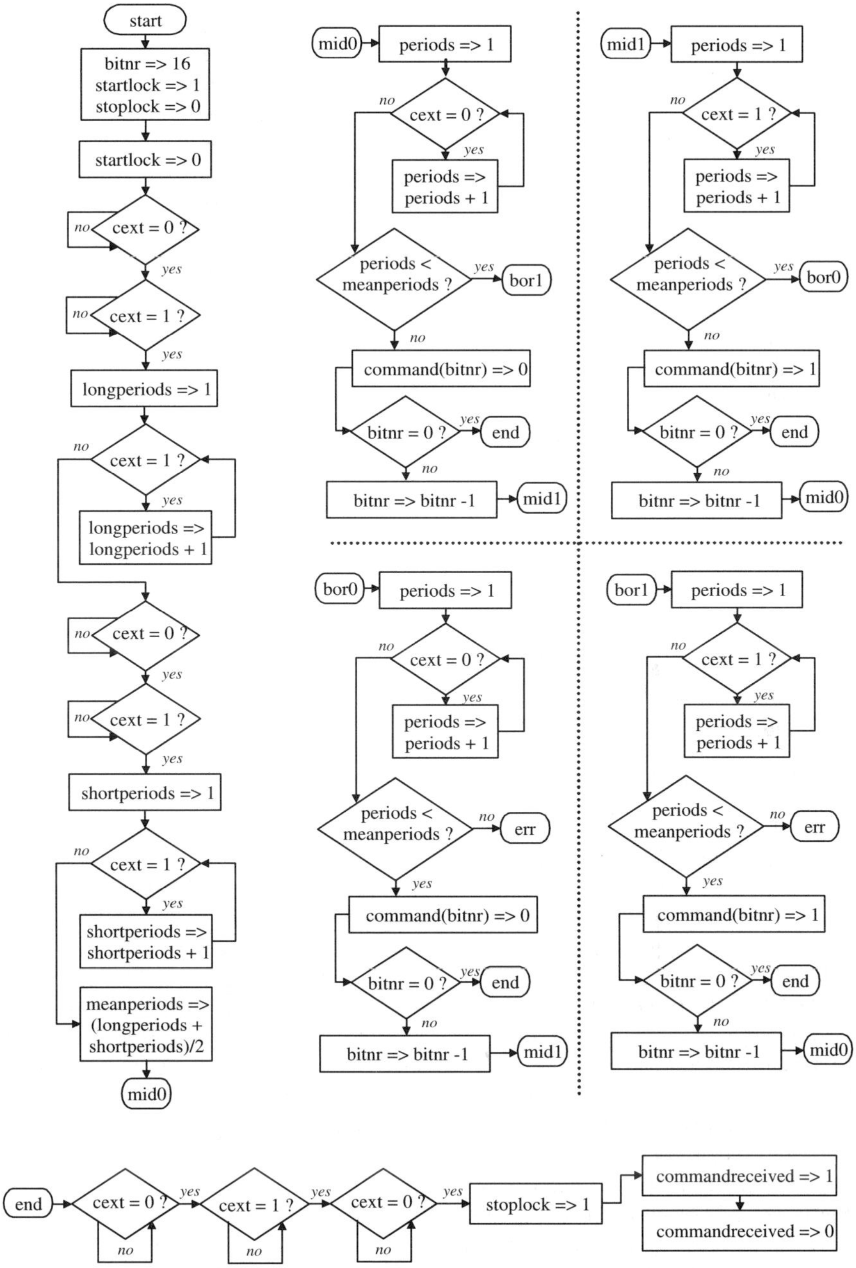

Figure 5.10: *Flowchart of the reception of a command.*

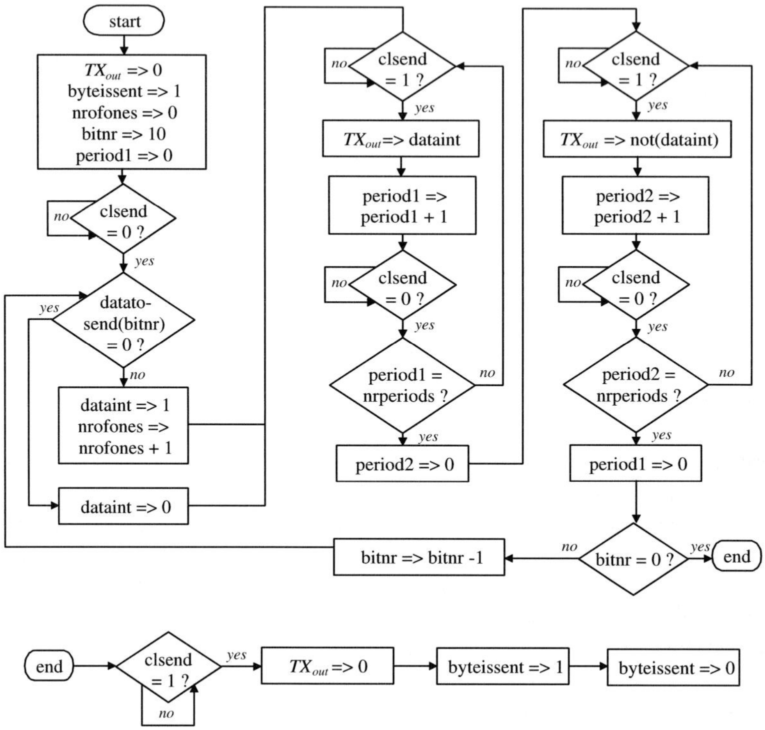

Figure 5.11: *Flowchart of the data transmission.*

The next building block is the transmission unit, which is used for both data and status bytes. The flowchart of the transmission/Manchester-encoding of the data/status bytes is shown in Fig. 5.11. A 3-bit header **101** is added to each byte that is transmitted, and both are combined into an 11-bit word *datatosend. bitnr* corresponds with the location of the bit of *datatosend* that is transmitted. After being triggered by the controller, the actual transmission is activated when a high-to-low transition of *clsend* (cf Section 5.4) occurs. After this activation *dataint* becomes equal to the MSB bit (*bitnr*=10) of *datatosend*, i.e. **1**. When the next low-to-high transition of *clsend* takes place (cf Section 5.4), TX_{out} becomes equal to *dataint* for a programmable number of periods of *clsend*, equal to *nrperiods*. This time interval corresponds with the transmission time of a half bit. After this period, TX_{out} becomes equal to the inverse of *dataint* when the next low-to-high transition of *clsend* takes place for *nrperiods* periods of *clsend*. In this way the transmitted bits are Manchester-encoded. This process is repeated until *bitnr* is equal to 0, i.e. for the LSB bit. The end of the transmission is signalled to the controller by the setting of *byteissent* when the next low-to-high transition of *clsend* appears. Each time a new bit is sent and if this bit is a 1, *nrofones* is increased with one. *nrofones* is equal to the number of ones

Data bits	Parameter
0	*amselect*
1	1
5-2	*seperationlength*
6	*fastcom*
7	0
8	*wireddata*
9	X

Table 5.3: Data bits of the transmission-mode command **01011.**

in *datatosend* at the end of the transmission. After the transmission of a full data page of 256 bytes or the transmission of the 21 status bytes, the remainder of the total number of ones in all the transmitted bytes divided by 256 is also sent. This extra parity byte is used to check for the correctness of the communication.

To program the number of periods *nrperiods* of *clsend*, corresponding with a half-bit of the Manchester-encoded bits, the command with command code **01001** is used. If this command is issued to the datalogger, *nrperiods* corresponds with the decimal value of the 4 LSB data bits incremented with one. The **01011**-command is also employed to configure the transmission mode. The meaning of the data bits of this command is illustrated in Table 5.3. The functions of *amselect* and *fastcom* have already been discussed before. Note that *amselect* must be set to 0 for the selected external transceiver [Avo], which can only interpret PM signals. *seperationlength* is a 4-bit word used to program the seperation period between the activation of the transmission and the actual transmission of the MSB bit of *datatosend*. During this period, which occurs before every transmitted byte, TX_{out} is reset. The total seperation period is equal to (*seperationlength*+1)·*nrperiods* periods of *clsend*. Note that this seperation period has not been included in Fig. 5.11. If the second MSB data bit *wireddata* is set, TX_{out} is buffered to a dedicated bondpad, so that it is possible to retrieve the data with a 'cable' for testing purposes.

The commands **01110** and **01101** are used to activate the transmission of a data page and the status bytes respectively. The data bits of the first command correspond with the binary representation of the number of the data page *datapage* that needs to be transmitted. Note that there is a difference between the sending of a data page and the sending of the status bytes. If the status bytes are transmitted, the external field must be switched off before the next status byte is sent, which is activated by switching the external field on again. Table 5.4 gives a list of the 21 status bytes and their meaning. Note that the combination of *durationon*, *coarsetimeron* and *processingalgorithm* corresponds with the 5 data bits in the right column of Table 5.2 for each algorithm.

During the reception of a command or the sending of data/status bytes a potential lock can occur. This can e.g. happen in the case where the external 132-kHz activation field falls away during the information exchange, and the datalogger waits for a clock transition of *cext*/*clsend* to occur (cf Fig. 5.10 and Fig. 5.11). To ensure that the datalogger is not stuck in the same

Status bytes	Parameter
statusbyte0(7)	*monitoringmodeon*
statusbyte0(6)	*badcommand*
statusbyte0(5)	*durationon*
statusbyte0(4)	*coarsetimeron*
statusbyte0(3-0)	*nrofdata(3-0)*
statusbyte1(7-5)	*processingalgorithm*
statusbyte1(4-0)	*strgnrrst(4-0)*
statusbyte2(7-3)	*strgnrlast(4-0)*
statusbyte2(2-0)	*nrperiods(3-1)*
statusbyte3(7)	*nrperiods(0)*
statusbyte3(6-0)	*upperthreshold(8-2)*
statusbyte4(7-6)	*upperthreshold(1-0)*
statusbyte4(5-0)	*lowerthreshold(8-3)*
statusbyte5(7-5)	*lowerthreshold(2-0)*
statusbyte5(4-0)	*coarsetimer(8-4)*
statusbyte6(7-4)	*coarsetimer(3-0)*
statusbyte6(3-0)	*datapage(9-6)*
statusbyte7(7-2)	*datapage(5-0)*
statusbyte7(1-0)	*samplecounter(17-16)*
statusbyte8(7-0)	*samplecounter(15-8)*
statusbyte9(7-0)	*samplecounter(7-0)*
statusbyte10(7-0)	*netimer end(7-0)*
statusbyte11(7-0)	*coarsetimerend(8-1)*
statusbyte12(7)	*coarsetimerend(0)*
statusbyte12(6-2)	*strgnrprog(4-0)*
statusbyte12(1-0)	*inputDAC(7-6)*
statusbyte13(7-2)	*inputDAC(5-0)*
statusbyte13(1-0)	*nullingwanted(8-7)*
statusbyte14(7-1)	*nullingwanted(6-0)*
statusbyte15(7-0)	*outputADC(8-1)*
statusbyte16(7-0)	*lockintreception(7-0)*
statusbyte17(7-0)	*lockintdatabytes(7-0)*
statusbyte18(7-0)	*lockintstatusbytes(7-0)*
statusbyte19(7-0)	obsolete
statusbyte20(7)	0
statusbyte20(6)	*fastcom*
statusbyte20(5-2)	*seplength*
statusbyte20(1)	1
statusbyte20(0)	*amselect*

Table 5.4: Overview of the status bytes.

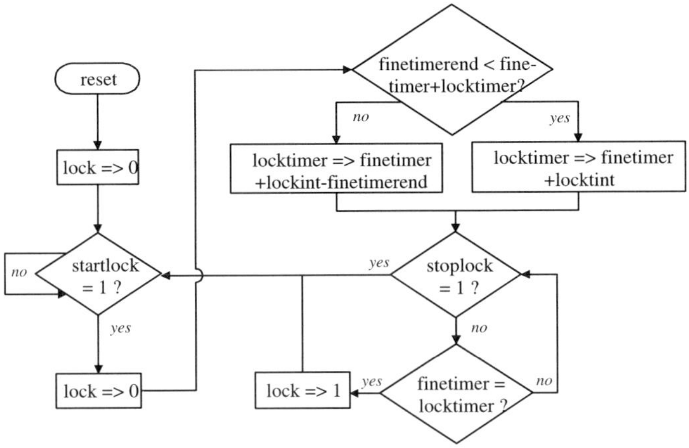

Figure 5.12: *Flowchart of the operation of the WDTs.*

MSB data bits	LSB data bits
1X	*lockintreception*
01	*lockintdatabytes*
00	*lockintstatusbytes*

Table 5.5: Meaning of the databits for the **01010**-command.

state 'forever' and that its normal operation is reassumed in this case, WDTs are implemented. Fig. 5.12 illustrates the operation of a WDT. At the reception/transmission start, the WDT is activated by setting *startlock* high. This results in the calculation of a threshold value *locktimer* for the fine timer, found by adding a programmable value *lockint* to the current value of *finetimer*. Hereby it is taken into account that *finetimer* is reset if an overflow occurs, which happens if *finetimer* becomes equal to *finetimerend*. If the end of the reception or transmission has not been signalled by the related units to the WDT by the setting of *stoplock* and *finetimer* becomes equal to *locktimer*, the WDT signals to the controller that a lock has occurred by setting *lock*. The controller takes care that the receiving and transmission units and the WDT are reset, and that the datalogger reassummes its normal operation and is able to communicate again with the external transceiver. If a lock occurs, the *badcommand* bit is also set.

To program the different values of *lockint* for the command reception and the data/status bytes transmission, the **01010**-command is used. Table 5.5 shows the role of the 8 LSB data bits, dependent on the 2 MSB data bits.

Figure 5.13: *Constellation E development board with an EPF10K130E Altera FPGA.*

5.7 Implementation and layout

The implementation/verification of the digital part including the internal transceiver has been carried out in several steps. First the building blocks have been written in VHDL, and simulated on the behavioral level seperately and together with the debugger included in SYNOPSYS. Next, the functionality of the digital part *together* with the sensor interface chip and the external transceiver, has been verified with a Constellation E development board [Nov], shown in Fig. 5.13. This board is equipped with an EPF10K130E Altera Field Programmable Gate Array (FPGA) and uses a PROM to program this. At power-up the data stored in the PROM are downloaded into the FPGA. The PROM is programmed with a ByteBlaster [Alt] connected to the parallel port of the PC by means of the programming software included in MAX+PLUSII. The conversion/compilation of the VHDL-code into the programming code for the FPGA is carried out by FPGACompilerII and MAX+PLUSII. The use of the development board has two important advantages: it allows fast prototyping of the code *and* it gives the possibility to test the complete digital part in combination with the the sensor interface chip and the external transceiver, restricting potential errors. The digital circuits have been implemented together with the sensor interface on the same chip, where SYNOPSYS has been used for synthesis, and APOLLO for the place-and-route of the standard cells. The latter took place at INVOMEC. The final simulation of the gate-level/standard-cell netlist has been performed with ModelSim.

Fig. 5.14 shows a photograph of the realized 0.7-μm CMOS datalogger IC, of which the area is equal to 12.8 x 4.6 mm^2. The complete digital part contains approximately 23400 gates and occupies 22.5 mm^2. A single power supply is used for the logic core and the digital input/output (IO) cells. Note that these IO cells are placed as far as possible from the analog part of the chip. Two IO cells in the top row differ from the other IO cells. These are the customly-designed output buffers driving the LC-tank TX (cf Fig. 5.3). Their transistor dimensions are chosen such that no current flows back from the LC-tank TX to the non-rechargeable batteries via these buffers during data transmission. The driving capability of the two buffers is different, i.e. 2.3

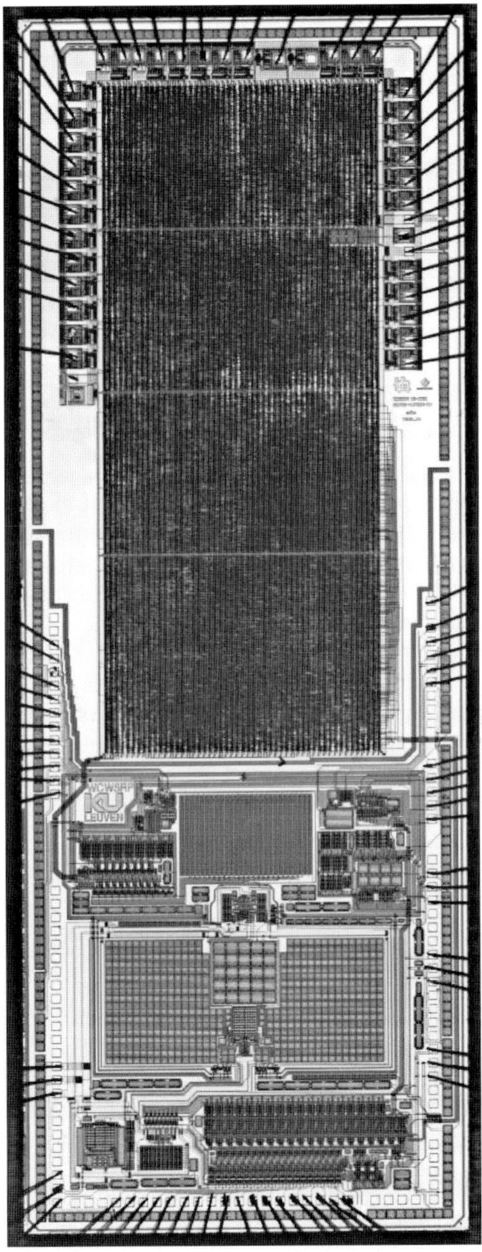

Figure 5.14: *Datalogger-IC photograph.*

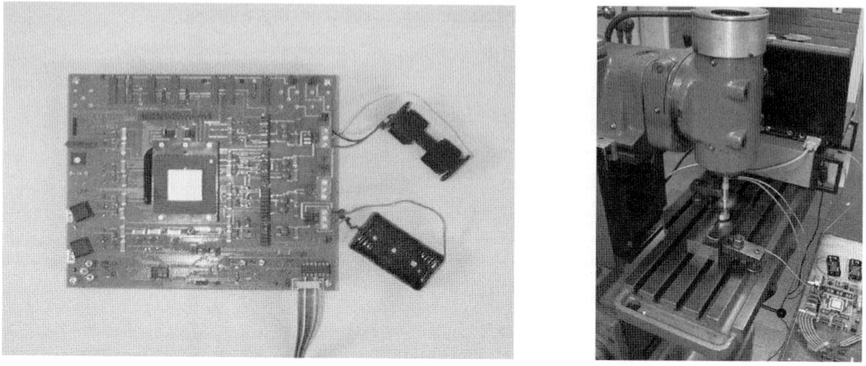

(a) (b)

Figure 5.15: *(a) Test PCB and (b) total test setup.*

mW and 4.1 mW respectively, which enlarges the flexibility of the datalogger. One of the two can be selected, depending on the expected distance between the datalogger and the external transceiver, by connecting a bondwire to the bondpad of the wanted buffer and the LC-tank TX. A Power-On-Reset (POR) has also been foreseen, so that the datalogger automatically is reset and takes on its normal operation after placement of the batteries.

To test the chip a dedicated test PCB has been developed. The datalogger IC, counting 95 IO pins in total, is packaged in a ceramic 145 Pin-Grid-Array (PGA) package. Like in the case of the sensor interface chip it is conductively glued to the grounded conductive base plane supporting the chip. The test PCB has a Zero-Insertion-Force (ZIF) PGA socket [3M], which is lever actuated. This allows to test different test chips on the same PCB. A photograph of the PCB, which has been designed with Cadstar, is shown in Fig. 5.15 (a).

5.8 Experimental results

The described test PCB has been used for the verification/testing of the datalogger. The measured maximum mean power consumption of the complete datalogger including the external SRAM during the monitoring mode is 136 μW per strain-gauge channel, which is lower than reported for similar systems [Ber 88, Fol 90, Cap 96, Beg 97]. Moreover, the presented datalogger has extra built-in intelligence. The power consumption of the digital part is approximatey equal to 150 μW. Elaborated testing of this part shows that it is fully functional. During data transmission the complete datalogger has a mean power consumption of 4.61 mW.

To perform both static and dynamic measurements a voltage-controlled piezoelectric actuator [Phy c], composed of PZT ceramic stacks, has been employed. A photograph of the total measurement setup is shown in Fig. 5.15 (b). Note that the actuator is clamped into a bench to obtain a very solid 'base' for it. The employed actuator is different from the one in the previous chapter. The new one has a displacement range of 90 μm (instead of 60 μm) with an accuracy

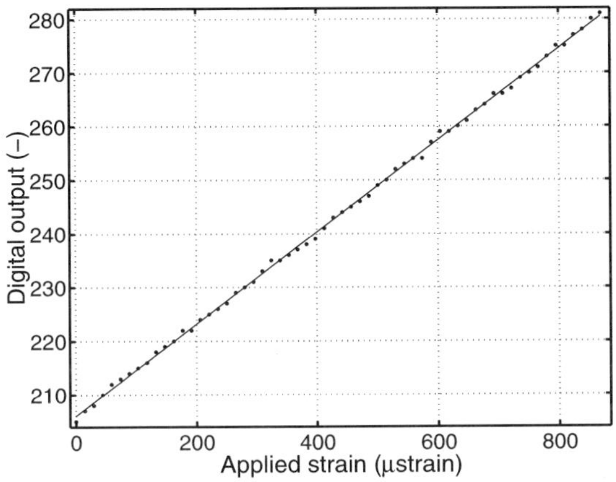

Figure 5.16: *Static measurement result. Dots: measurement data. Solid line: least-squares fit. σ < 6.5 μstrain.*

better than 0.2 μm. The maximum load that can be applied by the actuator is equal to 1000 N. A monitoring cell (not shown in Fig. 5.15 (b)) can be employed to measure the forces, imposed by the actuator. Fig. 5.16 gives the output data for a static measurement where an incrementally-increasing strain has been imposed to a strain gauge placed on a cylinder of rigid PVC (yielding larger strain values in comparison with titanium for the same maximum pushing force). The dots represent the measurement data and the solid line the least-squares fit. This measurement shows a standard deviation of the error smaller than 6.5 μstrain, proving that the static-measurement-accuracy requirement is satisfied. It can be seen from this measurement that the distribution of the errors is random, indicating that no systematic error is present anymore if this test setup is employed, in contrast with the measurement setup used for the sensor interface chip as explained in Section 4.5.5.

Two different measurement setups are employed for the dynamic measurements. The first one has already been discussed in Section 4.5.6. In the new measurement setup, shown in Fig. 5.17, the strain gauge is installed on a rigid-PVC beam supported at both sides. The beam is loaded by two point loads, symmetrically distributed around the center of the beam. The benefit of this approach is that the strain is constant between the two point loads [Roa 75, Rey], so that the alignment of the point loads and the strain gauge is not critical. The applied strain is proportional to the displacement of the piezo stack. Fig. 5.18 (a) shows a window of the measured output data for a sinusoidal strain with a peak-to-peak amplitude of 1005 μstrain and a 4-Hz frequency, which has been performed with this measurement setup. This amplitude corresponds with the maximum displacement that can be applied by the actuator. Fig. 5.18 (b) illustrates the PSD of the measured output data, yielding a SNDR of 35 dB, corresponding with a standard deviation σ of the error equal to 6.1 μstrain.

In the case of the measurement setup described in the previous chapter the PZT actuator

(a)

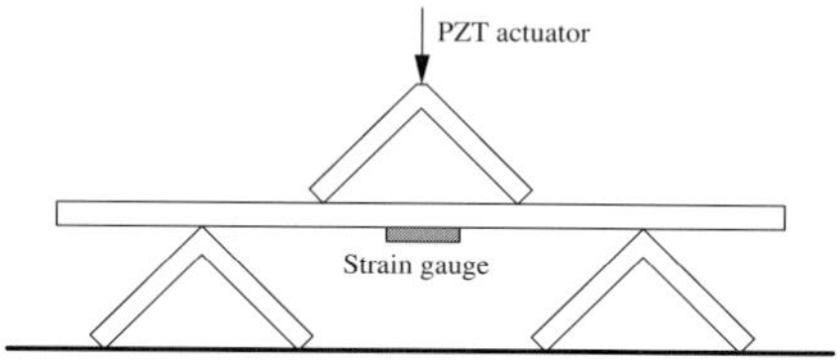

(b)

Figure 5.17: *(a) Photograph and (b) schematical representation of the measurement setup with a beam supported at both sides and loaded by two point loads.*

applies a displacement to a steel disc which is fixated by means of a M2 screw to an abutment. The weak connection due to the M2 screw and the bending of the steel disc itself introduce non-linearities in the measurements, resulting in a reduced linearity/accuracy of this test setup compared with the new measurement setup. Despite this drawback, this test setup allows to perform measurements with real abutments. An example of such a measurement with a real abutment is shown in Fig. 5.19. In this measurement a sinusoidal strain with a peak-to-peak amplitude of 479 μstrain and a 30-Hz frequency has been imposed. Note that in Fig. 5.19 (a) the data points are connected with each other so that it seems that the measured signal is modulated. This however is not the case and is due to the fact that the 30-Hz signal is close to the Nyquist frequency. The SNDR is equal to 27.2 dB, equivalent to a standard deviation σ of the error of 7 μstrain.

To investigate the influence of the integration of the digital part and the internal transceiver on the same chip as the sensor interface, the datalogger has been tested under similar conditions as the sensor interface chip (cf Section 4.5.6). This 2-Hz sinusoidal-strain measurement with a peak-to-peak amplitude of 514 μstrain yields a standard deviation σ of the error equal to 6.8 μstrain, which indicates that the accuracy degradation due to the single-chip integration of the datalogger IC is limited to about 0.6 μstrain, bearing in mind that the ADC problem is not present anymore. The limited accuracy degradation has been achieved by extensive (in-circuit) substrate contacts, guard rings, shielding, and, most importantly, by conductively glueing the datalogger-IC substrate to a grounded metallic plane. Also the influence of using a single supply instead of separate analog and digital supplies has been investigated. The accuracy degradation due the use of a single supply is limited to 0.3 μstrain, demonstrating that this influence is negligible.

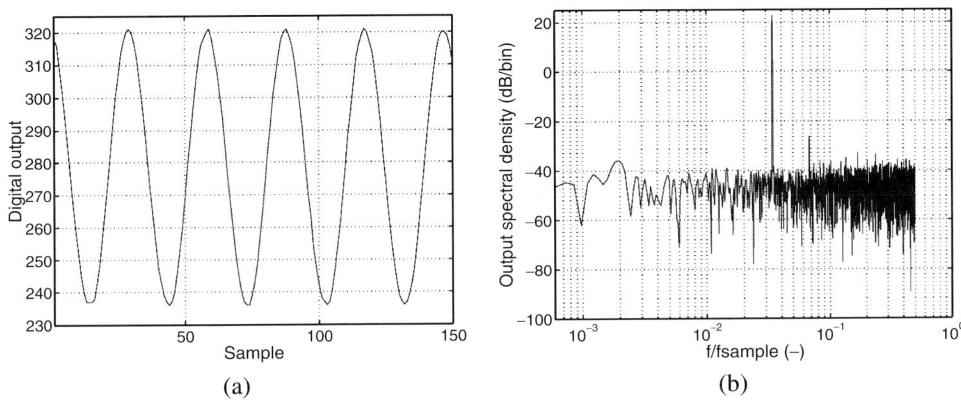

Figure 5.18: *(a) Output data and (b) PSD for a sinusoidal strain (f = 4 Hz, peak-to-peak amplitude = 1005 μstrain,* f_{sample} *= 118 Hz, SNDR = 35 dB).* σ *= 6.1 μstrain.*

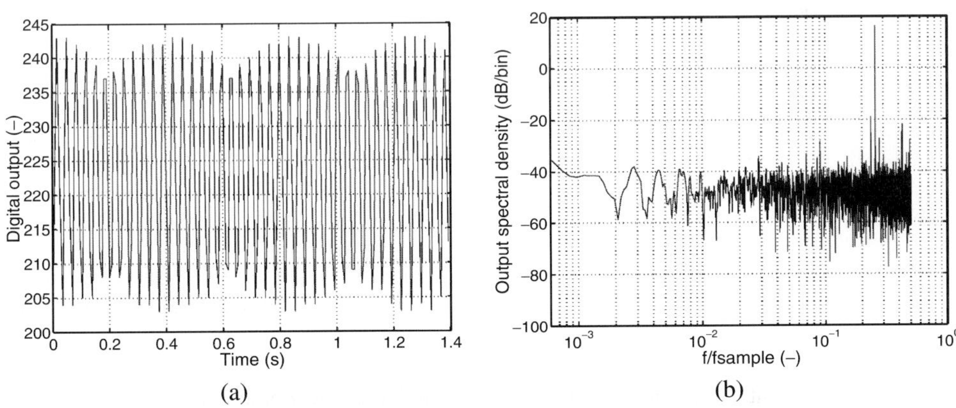

Figure 5.19: *(a) Output data and (b) PSD for a sinusoidal strain (f = 30 Hz, peak-to-peak amplitude = 479 μstrain,* f_{sample} *= 118 Hz, SNDR = 27.2 dB).* σ *= 7 μstrain.*

5.9 Future work: packaging

Fig. 5.20 shows the result of a first concept study, which has been done to find out how the datalogger, the ferrite-cored RF coils and the batteries can be embedded in an implant-retained dental prosthesis. Despite the fact that implant-supported fixed prostheses will be used in the final experiments instead of implant-retained prostheses (cf Section 3.2), the ideas gained during this study can be the starting point for the final packaging. Note that only 2 abutments are depicted in Fig. 5.20. The dental prosthesis is made of a synthetic material, which has two important advantages: a low shrinkage during molding and subsequent hardening, and a very low permeability for fluids. This allows to mold the electronics, placed on a carrier, directly in the prosthesis after being treated with an epoxy. During molding Teflon screws are put in the abutments to achieve correct positioning of the abutments in relation to the patient's oral model and to make sure that the inner space of the abutments and the space underneath and above remain free for the installation of the abutment screws after hardening. Teflon is utilized because of its high chemical inertness, so that no reaction with the synthetic prosthesis material occurs. The electronics are placed on a ceramic Al_2O_3 carrier, that contains at one side the 2-Mbit SRAM and at the other side the datalogger IC, the RF components and the resistors. The total volume of the electronics is approximately equal to twice the SRAM volume. To restrict possible interference the 24 sensitive connections, i.e. 18 strain-gauge-channel connections and one ground connection for each abutment, are placed at one side of the substrate. The connections to the battery holder and the receiver and transmitter coils L1/L2 are placed at the other side. A dedicated Teflon battery holder is employed to protect the batteries from the mouth environment, which is sealed by a Teflon cover during molding and hardening of the synthetic prosthesis material. The batteries may not be in contact with the prosthesis material during hardening, because the high temperature needed for the polymerisation of this material causes a capacity loss of the batteries. After hardening, the two batteries are placed in the holder and the top is sealed with a detachable cover, possibly imitating a tooth. This makes re-use of the system for a given patient possible after replacement of the batteries.

The final installation of the datalogger in an implant-supported prosthesis will be similar. However, because of the confined space in comparison with an implant-retained prosthesis, it is not possible to place all the components at one side. In the final design the datalogger probably will be placed centrally at the inside (oral side) of the prosthesis, which will be extended there if necessary. It is expected that this does not cause problems for the patient. The batteries and the RF parts will be placed at the right/left side of the datalogger. The final packaging and the design of the ceramic carrier need to be further elaborated in the future.

5.10 Conclusion

In this chapter the development of the single-chip intelligent-datalogger IC has been presented. It is composed of the multi-gauge sensor interface presented in the previous chapter, a digital part which adds intelligence, and a wireless bi-directional transceiver. The complete datalogger, which consists of this chip and an external SRAM memory, has two operation modes: a

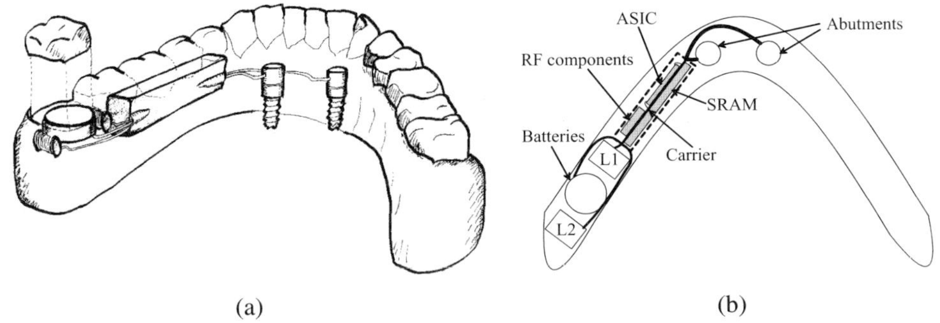

Figure 5.20: *(a) Drawing and (b) schematical representation of the installation of the datalogger in an implant-retained prosthesis.*

monitoring mode and a telemetry mode. In the former mode the sensor interface measures and digitizes the strain-gauge signals and the digital part processes the measured data. In the latter mode it is possible to reconfigure the datalogger and to collect stored data or status bytes from the datalogger by the bi-directional transceiver.

The different building blocks of the digital part have been discussed. Among other things it supervises the proper selection sequence of the strain-gauge channels and the storage of the processed 9-bit data in the 8-bit external SRAM. Moreover, it manages the in- and outgoing communication between the datalogger and the outside world. The communication protocol and the command-and-data formats have been presented. The digital part also contains an automatic compensation block which performs automatic nulling towards a user-definable output value for a selectable strain-gauge channel. To accomplish this a dedicated search algorithm is implemented, which determines the required compensation word by successive approximation. To reduce the required data storage capacity, a data processing unit with 8 selectable algorithms and adjustable parameters also has been incorporated. This unit ensures that only clinical relevant data are stored in the memory, and it allows to optimize the data processing towards a given patient/application, enlarging the flexibility of the datalogger.

To communicate with the datalogger a commercially available external transceiver has been adapted. The datalogger can be (re-)programmed by means of amplitude modulation of a 132-kHz programming field. On the other hand, data are transmitted from the datalogger to the external transceiver by modulation of a 66-kHz carrier, which is derived onboard of the datalogger from an incoming non-modulated 132-kHz field. In this way no on-chip oscillator is required to generate the transmission carrier. Moreover, the incoming 132-kHz field is externally controlled by a crystal oscillator, resulting in an improved communication. The transceiver is able to transmit data to the external transceiver over a distance of 30 cm at a maximum data rate of 4 kbytes/s with a mean power consumption of 2.3 mW. The list of commands that can be issued to the datalogger has been given. Also the meaning of the bits of the 21 status bytes has been presented.

Before the chip has been realized, the sensor interface chip has been tested together with the digital part and the transceiver by means of a FPGA. The datalogger IC has been fabricated in a 0.7-μm CMOS technology and measures 12.8 x 4.6 mm^2. Its digital part contains about 23400 gates and has a power consumption of approximately 150 μW. The measured maximum mean power consumption of the complete datalogger including the external SRAM during the monitoring mode is 136 μW per strain-gauge channel. During transmission of collected data/status bytes on the other hand the total power consumption is equal to 4.61 mW. Two different measurement setups have been used to test the datalogger chip: one with an abutment and one with a beam supported at both sides and loaded by two point loads. The former allows to perform measurements with a *real* abutment, while the latter has a superior linearity. The measured accuracy during static and dynamic measurements for both test setups is better than 10 μstrain, showing that the accuracy requirement is satisfied. To conclude, the result of a first concept study of the datalogger's packaging has been presented.

Chapter 6

Conclusion

In this work the development of an autonomous miniaturized intelligent datalogger for stress monitoring in oral implants has been presented. The device is employed to investigate the loads acting on oral implants supporting dental prostheses in order to gain more insight in the processes involved in bone remodeling and implant failures. The loads are measured by placing strain gauges on the abutments, which are positioned in the gums on top of the implants. By combining the measured resistance values of the 3 strain gauges installed on each abutment, the axial force and the bending moment, imposed on each abutment, can be derived. The datalogger is capable of monitoring up to 18 strain-gauge channels at a sample rate 111 Hz per channel. It is able to work autonomously over a period of two days, operated by two 1.55-V 41-mAh batteries.

The application of a miniaturized datalogger embedded within the dental prosthesis has several advantages. The measurements can be carried out without inconvenience for the patient in his/her normal living conditions, independent of the hospital, so that artificial chewing behavior is avoided. Moreover, since the device is able to monitor autonomously over a 2-day period, also unconscious nocturnal dental activities, seen as a missing link in the validation of existing bone remodeling models, can be monitored.

The datalogger system consists of 4 major parts, shown in Fig. 6.1: a sensor interface, a digital part and a bi-directional transceiver, integrated on a single-chip datalogger IC, and an external 2-Mbit SRAM memory. The ***sensor interface***, shown in Fig. 6.2, measures and digitizes the signals of the different strain-gauge channels. Because of the datalogger's complexity the sensor interface has been integrated first on a separate chip. It includes a reference current source, an 8-bit DAC, a digital interface and compensation-words memory, a SC amplifier, a SC S/H, a 9-bit successive approximation ADC and a 6-bit programmable relaxation clock oscillator. A current-driven Wheatstone configuration is implemented to interface with the strain gauges resulting in a lower power consumption compared to a voltage-driven one. The offsets introduced in the different channels due to pre-strains, current-mirror inaccuracies and the strain gauges' resistance tolerance can be digitally compensated by the inclusion of the DAC, the digital interface and the compensation-words memory. The amplifier, S/H, and the preamplifier of the successive approximation ADC all include 1/f-noise and offset (drift) cancellation, based on the CDS technique. The measured mean current consumption of the sensor interface per strain-gauge channel is lower than 40 μA and its dynamic accuracy is better than 10 μstrain.

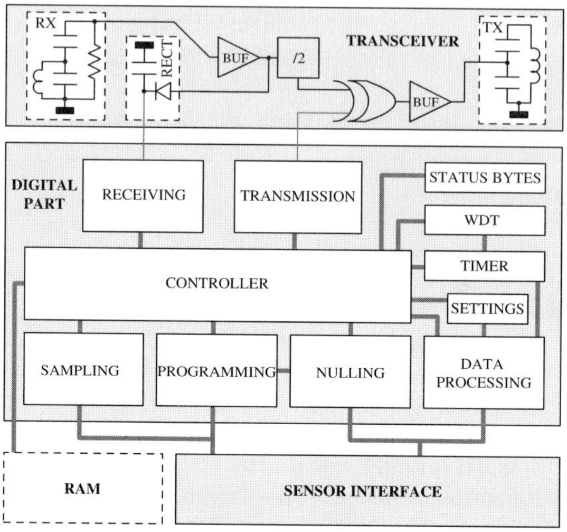

Figure 6.1: *Overview of the complete datalogger system.*

The ***digital part*** adds intelligence to the datalogger. It orchestrates the operation of the device and supervises the strain-gauge selection sequence, the data storage and the in- and outgoing communication. It is implemented by means of a custom-designed 23.4-kgates FSM. The power consumption of the FSM is limited to 150 μW. Two important features are implemented, enlarging the datalogger's flexibility:

- An automatic-compensation block to perform automatic nulling towards a user-definable output value for a selectable strain-gauge channel.
- A programmable data processing unit with 8 selectable algorithms and adjustable parameters, which reduces the required data storage capacity. It ensures that only clinical relevant data are stored in the memory and makes optimization of the data processing towards the patient/application possible.

The bi-directional ***transceiver*** allows to wirelessly retrieve collected data or status bytes from the datalogger *and* to reconfigure the measurement device *in situ*, which increases its flexibility. In this way the offsets introduced in the different channels can be compensated for *after* placement of the prosthesis. The integrated transceiver is able to communicate over a distance of 30 cm at a data rate of 4 kbytes/s with a mean power consumption of 2.3 mW.

The intelligent-datalogger IC has been fabricated in a 0.7-μm CMOS technology. The maximum measured mean power consumption of the complete datalogger, consisting of this chip and the external SRAM, in its monitoring mode is restricted to 136 μW per strain-gauge channel. Measurements also have shown that the datalogger satisfies the accuracy requirement of 10 μstrain with an ample margin.

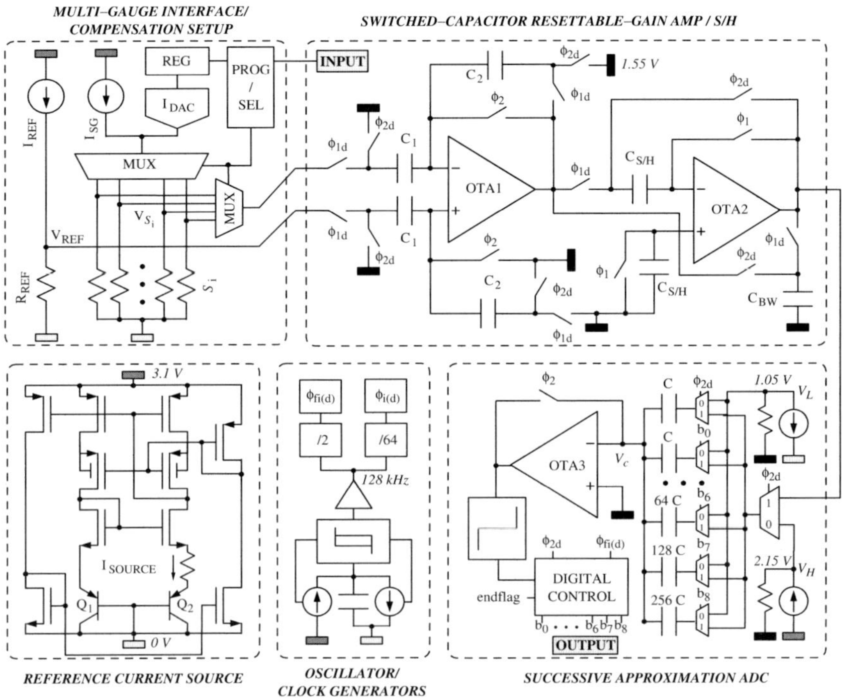

Figure 6.2: *Schematic overview of the sensor interface.*

In comparison with comparable state-of-the-art strain-gauge measurement systems, like e.g. [Ber 88, Fol 90, Cap 96, Beg 97], this work has the following strong points and novelties:

- The mean power consumption of the complete datalogger during monitoring is restricted to 136 μW/channel, which is to the author's knowledge the lowest ever presented for comparable measurement systems.
- The single-chip datalogger IC combines a sensor interface, a digital unit and a transceiver. Only a RAM and external transceiver components are needed to realize a complete autonomous wireless datalogger with a high degree of flexibility.
- The integrated transceiver allows programmation of the datalogger's operation mode *in situ* after placement of the prosthesis.
- The strain-gauge datalogger IC contains an onboard wireless-programmable data processing unit, which allows to fully optimize the processing algorithm for each individual patient. Only clinical relevant data are stored in the RAM reducing the necessary data storage capacity drastically.

- The datalogger is capable to compensate itself for the offsets introduced in the strain-gauge channels.
- The integrated sensor interface includes a current reference, an 8-bit DAC, a digital interface and nulling memory, an offset-cancelled SC amplifier, an offset-cancelled SC SH, a 9-bit successive approximation ADC and a 6-bit programmable relaxation clock oscillator.
- Contrary to the existing external system the measurements are no longer restricted to the hospital environment and in addition unconscious nocturnal dental activities can also be monitored. Moreover, artificial chewing behaviour is avoided due to the absence of strain-gauge wires.

This work has demonstrated the feasibility of the single-chip integration of an intelligent strain-gauge datalogger IC, combining a sensor interface with digitally-programmable offset-compensa-tion, a digital unit with adjustable data-processing and automatic offset-compensation, and a wireless bi-directional transceiver. The concept of the presented intelligent datalogger is not restricted to dental prostheses only. Due to its versatility, it can be applied in different kinds of portable personal health monitoring systems. Therefore this work hopes to be a -albeit small-step forward in miniaturized intelligent dataloggers from which all of mankind may benefit one day.

Appendix A

Transistor dimensions

Transistor	W (μm)	L (μm)
M1	104	13
M2	104	13
M3	31.6	1.2
M4	31.6	1.2
M5	151.6	14
M6	151.6	14
M7	4.3	10
M8	2.4	43
M9	2.4	43
M10	2.2	15.1
M11	4.5	0.7
M12	22.5	1.2
M13	4.5	0.7
M14	22.5	1.2
Capacitor	**C (pF)**	
C_{START}	20	
C_{DEL}	20	

Table A.1: Dimensions of the components in Fig. 4.4.

Transistor	W (μm)	L (μm)
$M_{cur,new}$	10.6	31
$M_{cas,new}$	2.2	1.2
M	12.4	1.2
SW,PMOS	12.8	1.2
SW,NMOS	2.6	0.7

Table A.2: Dimensions of the transistors in Fig. 4.13.

Transistor	W (μm)	L (μm)
M1	104	13
M2	31.6	1.2
M3	26	12
M4	7	25
M5	7	25
M6	26	12
M7	2.2	1.2
M8	10.6	31
M10	2.2	35.9
M11	3.2	10
M12	3.2	10
M13	2.2	120.2
M20	2.2	8
M21	2.2	39.6
M22	2.2	39.6
M23	2.2	21.7

Table A.3: Dimensions of the transistors in Fig. 4.15.

Transistor	W (μm)	L (μm)
M1a	54.5	5
M1b	54.5	5
M2	31.2	1.2
M3	31.2	1.2
M4	12.4	1.2
M5	12.4	1.2
M6	72.4	17
M7	72.4	17
M8	929	17
M9	929	17
M10	144	17
M11	2.2	11.2
M12	2.3	11
M13	2.3	11
M14	2.2	73.5

Table A.4: Dimensions of the transistors in Fig. 4.30.

Transistor	W (μm)	L (μm)
M1a	119	4
M2a	119	4
M3a	25.3	6
M4a	15.5	14
M5a	15.5	14
Moff	42	0.7
M1c	18	5.8
M2c	18	5.8
M3c	25.3	6
M4	9	5.8
M5	9	5.8
M6	22.5	5.8
M7	22.5	5.8
M8	3	5.8
M9	3	5.8
M10	7.5	5.8
M11	7.5	5.8
M12	9	5.8

Table A.5: Dimensions of the transistors in Fig. 4.42.

Transistor	W (μm)	L (μm)
M1	36.5	14
M2	36.5	14
M3	87.5	14
M4	87.5	14
M5	87.5	14
M6	87.5	14
M7	7	14
M8	7	14
M9	16.8	14
M10	36.5	14
M11	30.5	5
M12	25.3	6

Table A.6: Dimensions of the transistors in Fig. 4.43.

Transistor	W (μm)	L (μm)
Mn	2.6	35
Mp	4	23
M1	2.2	4
M2	12.3	1.2
M3	12.3	1.2
M4	2.2	5.2
M5	2.5	1.2
M6	2.5	1.2
Mres	4.5	0.7
SW_1,PMOS	7.6	0.7
SW_1,NMOS	2.2	0.7
SW_2,PMOS	130	0.7
SW_2,NMOS	40	0.7
Capacitor	**C (pF)**	
Cosc	5.1	

Table A.7: Dimensions of the transistors in Fig. 4.46.

Transistor	W (μm)	L (μm)
M1	2.6	35
M2	2.6	35
M3	3.7	14
M4	3.7	14
M5	4	23
M6	4	23
M7	2.6	35
M8	2.6	35
M9	4	23
Resistor	**R (MΩ)**	
$R_{CL,int}$	3.22	
$R_{CL,ext}$	1.97	

Table A.8: Dimensions of the components in Fig. 4.47.

Appendix B

Digital error correction of ADC

For completeness this appendix presents the (digital) error correction algorithm, which has been employed for the ADC of the sensor interface chip. Since the ADC problem has been solved in the final datalogger chip, this correction algorithm, which also has been implemented in the final chip, is obsolete though. Note that the parameters of the correction algorithm on board of the datalogger are programmable by the wireless link.

Detailed analysis of the measurement results of the sensor-interface-chip ADC showed that the same pattern of missing codes appears for every chip, indicating a systematic error. Digital output codes are missing at the multiples of 32: 4 codes are missing at 256, 2 codes at 128 and 384, and 1 code at the other multiples of 32. To deal with the missing codes the correction algorithm, listed below, is used, offering a solution for this particular case. The algorithm calculates correction terms (*corr1, corr2, corr3 and corr4*) dependent on the interval to which the measured ADC output *ADvalue* belongs. These correction terms and a fixed term (+10) are added to *ADvalue*, yielding the corrected value *correctAD*. The fixed term is used to center the resulting corrected codes around 256. The algorithm transforms the missing codes to the outside borders of the range (0-511). Note that the errors are not solved by this algorithm, but that their effect is reduced in the range of interest. To solve the ADC problem the layout of its capacitor array has been changed.

Error correction algorithm

```
function [correctAD] = correctAD(ADvalue)

intermediate=ADvalue;

% Correction term 1:
if intermediate <256 corr1=0;
elseif intermediate >=256 corr1=-4;
end;

% Correction term 2:
if intermediate <128 corr2=0;
elseif intermediate >=128 & intermediate <384 corr2=-2;
elseif intermediate >=384 corr2=-4;
end;
```

```
% Correction term 3:
if intermediate <64 corr3=0;
elseif intermediate >=64 & intermediate <192 corr3=-1;
elseif intermediate >=192 & intermediate <320 corr3=-2;
elseif intermediate >=320 & intermediate <448 corr3=-3;
elseif intermediate >=448 corr3=-4;
end;

% Correction term 4:
if intermediate <32 corr4=0;
elseif intermediate >=32 & intermediate <96 corr4=-1;
elseif intermediate >=96 & intermediate <160 corr4=-2;
elseif intermediate >=160 & intermediate <224 corr4=-3;
elseif intermediate >=224 & intermediate <288 corr4=-4;
elseif intermediate >=288 & intermediate <352 corr4=-5;
elseif intermediate >=352 & intermediate <416 corr4=-6;
elseif intermediate >=416 & intermediate <480 corr4=-7;
elseif intermediate >=480 corr4=-8;
end;

correctAD=intermediate+corr1+corr2+corr3+corr4+10;
```

Appendix C

Sampling unit

C.1 VHDL code

```
-- Sampling unit --

library WORK;
library IEEE;
use IEEE.STD_LOGIC_1164.all;
use IEEE.STD_LOGIC_arith.all;
use WORK.FUNCPROC.all;                -- Package with functions/procedures
                                      -- and type/constant definitions.

entity sampling is
  port(
    -- Clock:
    samplingmodeclock   : in std_ulogic; -- Clock of the sampling unit: 128 kHz.

    -- Inputs:
    resetsamplingmode   : in std_ulogic; -- Reset pin.
    periodcounteron     : in std_ulogic; -- Flag to indicate that the periodcounter
                                         -- needs to be memorized.
    timeron             : in std_ulogic; -- Flag to indicate that the (coarse) timer
                                         -- needs to be memorized.
    samplingmodeon      : in std_ulogic; -- Flag to indicate that sampling is
                                         -- going on. Sampling can be activated by
                                         -- the startsampling-command and can be
                                         -- stopped by the stopsampling-command.
    strgnrfirst         : in std_ulogic_vector (4 downto 0); -- First strain gauge
                                         -- to be processed: programmable register.
    strgnrlast          : in std_ulogic_vector (4 downto 0); -- Last strain gauge
                                         -- to be processed: programmble register.
                                         -- strgnrfirst and strgnrlast determine
                                         -- the number of selected strain-gauge channels.
    LDsamplecounter     : in std_ulogic; -- Flag used to indicate that the
                                         -- samplecounter has been reprogrammed
                                         -- by the main controller.
    endflag             : in  std_ulogic; -- Flag to signal the end of an AD-conversion.
    dataready           : in  std_ulogic; -- Flag to indicate that the data
                                         -- are ready to be written into the
                                         -- memory. This flag has to go to zero
                                         -- before the next memory write can be
                                         -- carried out.
    strgnrthreshold     : in std_ulogic_vector (4 downto 0); -- Used for correct
                                         -- memorizing of the data. Corresponds
                                         -- with the strain-gauge no of the last
                                         -- strain-gauge channel of a data set that
                                         -- needs to be memorized. strgnrthreshold is
                                         -- controlled by the data processing unit.
    lastperiodcounter   : in  std_ulogic_vector(8 downto 0); -- Used to count the
                                         -- periods between two threshold crossings.
    timerdata           : in  std_ulogic_vector(8 downto 0); -- Coarse timer output.
    strgnrmemory        : in  std_ulogic_vector (4 downto 0); -- Strain-gauge channel no
                                         -- of the data to be put in the memory when data
                                         -- are ready (i.e. when dataready goes high).
```

```
    processeddata         : in   std_ulogic_vector (8 downto 0); -- Data, processed by the
                                                -- data procesing unit, to be written to
                                                -- the memory.
    valuesamplecounter : in   memoryaddress -- Programmable value of the global
                                                -- memory location where the next
                                                -- data set has to be written in the memory.

    -- Outputs:
    strgnrcurrent              : out std_ulogic_vector (4 downto 0); -- Current strain-
                                                -- gauge no of the data at the ADC output.
                                                -- There is a delay between the selection
                                                -- of a certain strain gauge (DAC input) and its
                                                -- ADC output.
    strgnrinputout             : out std_ulogic_vector (4 downto 0); -- Strain-gauge no at
                                                -- the DAC input. The strain gauge has to be
                                                -- selected when endflag = 0.
    processingonout            : out std_ulogic; -- Flag to indicate that data processing
                                                -- is going on. The processed data have to be
                                                -- written to the memory. Data processing (by the
                                                -- processing unit) only starts when the data of the
                                                -- first selected strain-gauge channel are
                                                -- available at the ADC output.
    memorypointerout           : out memoryaddress; -- Local memory location where the
                                                -- next (strain-gauge) data have to be
                                                -- written in the memory.
    memorydata                 : out std_ulogic_vector (7 downto 0); -- Data to be
                                                -- written to the memory.
    datamemorizingout          : out std_ulogic; -- Flag to indicate that data memorizing
                                                -- is going on. This flag is also used
                                                -- to pass the control to the main controller.
    newsamplingmodecycleout : out std_ulogic; -- Flag used to indicate that a
                                                -- new sampling-mode cycle can start and to
                                                -- pass the control to the main controller.
    samplecounterout           : out memoryaddress; -- Global memory location where the next
                                                -- data set has to be written in the memory.
    CE2                        : out std_ulogic; -- Chip enable input of memory. CE2 = 1 when
                                                -- read/write operation takes place, otherwise 0.
    );

end sampling;

architecture behaviour of sampling is
  type statesamplingmode is (sampmodestart, sampmode1, sampmode2, sampmode3, sampmode4a,
  sampmode4b, sampmode4c, sampmode5, sampmode6, sampmode6b, sampmode7a, sampmode7b, sampmode8,
  sampmodeend1, sampmodeend2, sampmodeend3, storetimer1, storetimer2, storetimer3,
  storeperiodcounter1, storeperiodcounter2, storeperiodcounter3);
  signal samplingmodestate : statesamplingmode;
  signal strgnrprevious : std_ulogic_vector (4 downto 0);
  signal memend                        : memoryaddress; -- Memory location where
                                                -- the lastbits-byte is stored after
                                                -- stopping the sampling mode.
  signal samplecountertimer            : memoryaddress;
  signal samplecounterperiodcounter : memoryaddress;
  signal lastbits                      : std_ulogic_vector (7 downto 0); -- Byte with the LSB
                                                -- bits (9-bit data words) of the last processed-
                                                -- data words.
  signal lastbit                       : std_ulogic; -- LSB bit of the last (9-bit) processed-
                                                -- data word.
  signal lastbitflag                   : std_ulogic; -- Flag to indicate that the lastbits-byte
                                                -- still has to be memorized.
  signal storetimer                    : std_ulogic; -- Flag to indicate that the timer
                                                -- needs to be stored.
  signal storeperiodcounter            : std_ulogic; -- Flag to indicate that the periodcounter
                                                -- needs to be stored.
  signal strgnrinput                   : std_ulogic_vector (4 downto 0);
  signal memorypointer                 : memoryaddress;
  signal newsamplingmodecycle          : std_ulogic;
  signal datamemorizing : std_ulogic;
  signal samplecounter   : memoryaddress;
  signal processingon    : std_ulogic;

begin
  processingonout          <= processingon;
  samplecounterout         <= samplecounter;
  datamemorizingout        <= datamemorizing;
  strgnrinputout           <= strgnrinput;
```

```
memorypointerout          <= memorypointer;
newsamplingmodecycleout <= newsamplingmodecycle;

datasamplingmode : process(samplingmodeon, samplingmodeclock)

begin

  if (samplingmodeclock'event and samplingmodeclock = '1') then

    if (resetsamplingmode = '1') then

      CE2                    <= '0';
      processingon           <= '0';
      samplingmodestate      <= sampmodeend3;
      newsamplingmodecycle <= '1';
      datamemorizing         <= '0';
      lastbitflag            <= '0';
      samplecounter          <= valuesamplecounter;
      strgnrinput            <= strgnrfirst;

    else

      if (samplingmodeon = '1') then

        if (newsamplingmodecycle = '1') then
          newsamplingmodecycle <= '0';
          datamemorizing       <= '0';
          samplingmodestate    <= sampmodestart;
          storetimer           <= '0';
          storeperiodcounter   <= '0';
        end if;

        case samplingmodestate is
          when sampmodestart =>
            if (endflag = '1') then
              samplingmodestate <= sampmode1;
            end if;

          when sampmode1 =>
            if (endflag = '0') then
              strgnrinput       <= strgnrfirst;
              samplingmodestate <= sampmode2;
            end if;

          when sampmode2 =>
            if (endflag = '1') then
              strgnrprevious    <= strgnrinput;
              samplingmodestate <= sampmode3;
            end if;

          when sampmode3 =>
            if (endflag = '0') then
              if (strgnrinput = ('1', '0', '0', '0', '1')) then  -- Pointer has to
                                                                   -- go around in 18
                                                                   -- * 1/2000s = 9ms.
                strgnrinput      <= ('0', '0', '0', '0', '0');
              else
                strgnrinput      <= nat_to_5bits(bits5_to_nat(strgnrinput)+1);
              end if;
              processingon      <= '1';
              strgnrcurrent     <= strgnrprevious;
              samplingmodestate <= sampmode4a;
            end if;

          when sampmode4a =>
            if (endflag = '1') then
              samplingmodestate <= sampmode4b;
            end if;

          when sampmode4b =>
            if (dataready = '1') then
              memorypointer     <= nat_to_18bits(bits18_to_nat(samplecounter)+bits5_to_nat(
              strgnrmemory)-bits5_to_nat(strgnrfirst));
              memend            <= memorypointer;
              memorydata        <= processeddata(8 downto 1);
              lastbit           <= processeddata(0);
              datamemorizing    <= '1';
              samplingmodestate <= sampmode5;
            else
```

```
      samplingmodestate <= sampmode4c;
    end if;
    strgnrprevious         <= strgnrinput;

  when sampmode4c =>
    if (dataready = '1') then
      memorypointer   <= nat_to_18bits(bits18_to_nat(samplecounter)+bits5_to_nat(
      strgnrmemory)-bits5_to_nat(strgnrfirst));
      memend           <= memorypointer;
      memorydata       <= processeddata(8 downto 1);
      lastbit          <= processeddata(0);
      datamemorizing   <= '1';
    end if;
    samplingmodestate <= sampmode5;

  when sampmode5 =>
    if (endflag = '0') then
      if (strgnrinput = ('1', '0', '0', '0', '1')) then
        strgnrinput <= ('0', '0', '0', '0', '0');
      else
        strgnrinput <= nat_to_5bits(bits5_to_nat(strgnrinput)+1);
      end if;
      strgnrcurrent <= strgnrprevious;
      if (datamemorizing = '1') then
        -- Timing information is stored when stgnrmemory = strgnrlast
        -- (for some of the processing algorithms). The samplecounter
        -- calculation is different when timing information needs to be
        -- stored: one extra memory location is needed for the timing info.
        if (strgnrmemory = strgnrlast) then
          if (timeron = '1') then
            storetimer           <= '1';
            samplecountertimer <= samplecounter;
          end if;
          if (periodcounteron = '1') then
            storeperiodcounter          <= '1';
            samplecounterperiodcounter <= samplecounter;
          end if;
        end if;
        if (strgnrmemory = strgnrthreshold) then
          if (periodcounteron = '1') then
            if (bits18_to_nat(samplecounter) < overloadRAM) then
              samplecounter    <= nat_to_18bits(bits18_to_nat(samplecounter)+
              bits5_to_nat(strgnrlast)-bits5_to_nat(strgnrfirst)+3);
            end if;
            -- One extra memory location after each data set is required
            -- for both the timer and the periodcounter.
          else
            if (timeron = '1') then
              if (bits18_to_nat(samplecounter) < overloadRAM) then
                samplecounter <= nat_to_18bits(bits18_to_nat(samplecounter)+
                bits5_to_nat(strgnrlast)-bits5_to_nat(strgnrfirst)+2);
              end if;
            else
              if (bits18_to_nat(samplecounter) < overloadRAM) then
                samplecounter <= nat_to_18bits(bits18_to_nat(samplecounter)+
                bits5_to_nat(strgnrlast)-bits5_to_nat(strgnrfirst)+1);
              end if;
            end if;
          end if;
        end if;
        CE2              <= '1'; -- Chip is enabled => data are written
                                 -- to the memory.
      end if;
      samplingmodestate <= sampmode6;
    end if;

  -----------------------
  -- LSB bit processing: -
  -----------------------

  when sampmode6 =>          -- Further processing of the LSB bits of
                             -- the 9-bit data words.
    CE2               <= '0';
    samplingmodestate <= sampmode6b;

  when sampmode6b =>
```

```
    if (datamemorizing = '1') then
      lastbits(bits3_to_nat(memorypointer(2 downto 0))) <= lastbit;
      if (bits3_to_nat(memorypointer(2 downto 0)) = 7) then  -- LSB bits are written
                                                              -- to the memory when
                                                              -- the lastbits-byte
                                                              -- is full.
        memorypointer                                 <= nat_to_18bits(
        bits15_to_nat(memorypointer(17 downto 3))+lastbitsbegin);
        lastbitflag                                   <= '1';
      end if;
    end if;
    samplingmodestate                                 <= sampmode7a;

  when sampmode7a =>
    if (datamemorizing = '1') then
      if (lastbitflag = '1') then
        memorydata      <= lastbits;
      end if;
    end if;
    samplingmodestate  <= sampmode7b;

  when sampmode7b =>
    if (datamemorizing = '1') then
      if (lastbitflag = '1') then
        CE2             <= '1';
      end if;
    end if;
    samplingmodestate  <= sampmode8;

  when sampmode8 =>
    if (datamemorizing = '1') then
      if ((storetimer/='1') and (storeperiodcounter/='1')) then
        datamemorizing <= '0';
      end if;
    end if;
    CE2                 <= '0';  -- Data have been written to
                                 -- the memory.
                                 -- Chip is disabled => no
                                 -- further data can be written
                                 -- to the memory.
    lastbitflag         <= '0';

    if (storetimer = '1') then
      samplingmodestate    <= storetimer1;
    else
      if (storeperiodcounter = '1') then
        samplingmodestate <= storeperiodcounter1;
      else
        samplingmodestate <= sampmode4a;
      end if;

    end if;

    -------------------------
    -- Storing of the timer: -
    -------------------------

  when storetimer1 =>
    memorydata          <= timerdata(8 downto 1);
    lastbit             <= timerdata(0);
    memorypointer       <= nat_to_18bits(bits18_to_nat(samplecountertimer)+bits5_to_nat
    (strgnrlast)-bits5_to_nat(strgnrfirst)+1);
    memend              <= memorypointer;
    samplingmodestate  <= storetimer2;

  when storetimer2 =>
    CE2                 <= '1';
    samplingmodestate  <= storetimer3;

  when storetimer3 =>
    CE2                 <= '0';
    storetimer          <= '0';
    samplingmodestate  <= sampmode6;

    ---------------------------------
    -- Storing of the periodcounter: -
    ---------------------------------
```

```
          when storeperiodcounter1 =>
            memorydata          <= lastperiodcounter(8 downto 1);
            lastbit             <= lastperiodcounter(0);
            memorypointer       <= nat_to_18bits(bits18_to_nat(samplecounterperiodcounter)+
            bits5_to_nat(strgnrlast)-bits5_to_nat(strgnrfirst)+2);
            memend              <= memorypointer;
            samplingmodestate <= storeperiodcounter2;

          when storeperiodcounter2 =>
            CE2                 <= '1';
            samplingmodestate <= storeperiodcounter3;

          when storeperiodcounter3 =>
            CE2                 <= '0';
            storeperiodcounter <= '0';
            samplingmodestate   <= sampmode6;

          when others => null;
        end case;

      else

        if (processingon = '1') then    -- If processingon = 1 and the sampling
                                        -- mode has been stopped already, the
                                        -- lastbits-byte is written first to
                                        -- the memory before the main
                                        -- controller takes over the control.
          processingon <= '0';
          samplingmodestate <= sampmodeend1;
        else
          samplingmodestate <= sampmodeend3;

        end if;

        case samplingmodestate is

          when sampmodeend1 =>
            memorypointer       <= nat_to_18bits(bits15_to_nat(memend(17 downto 3))+
            lastbitsbegin);
            memorydata          <= lastbits;
            samplingmodestate <= sampmodeend2;

          when sampmodeend2 =>
            CE2                 <= '1';
            samplingmodestate <= sampmodeend3;

          when sampmodeend3 =>
            CE2                 <= '0';
            samplingmodestate <= sampmodeend3;
            newsamplingmodecycle <= '1';
            lastbitflag         <= '0';
            if (LDsamplecounter = '1') then
              samplecounter       <= valuesamplecounter;
            end if;

          when others => null;

        end case;

      end if;

    end if;

  end if;

end process;

end behaviour;
```

C.2 Flowchart

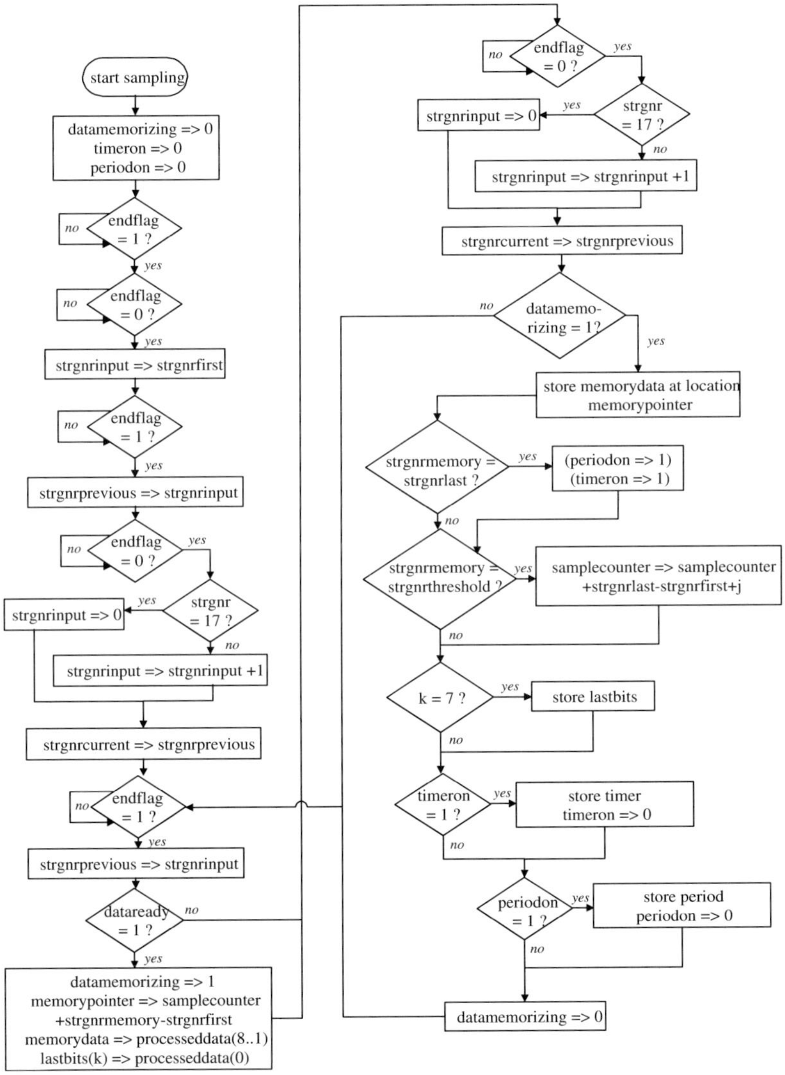

Figure C.1: *Simplified flowchart of the sampling unit. Note that k is equivalent with the decimal representation of the 3 LSB bits of memorypointer and that j depends on the selected algorithm.*

List of Publications

Journal papers

1. W. Claes, M. De Cooman, W. Sansen and R. Puers, "A 136-μW/channel autonomous strain-gauge datalogger", *IEEE Journal of Solid-State Circuits*, accepted for publication.

2. W. Claes, R. Puers, W. Sansen, M. De Cooman, J. Duyck and I. Naert, "A low power miniaturized autonomous data logger for dental implants", *Sensors and Actuators*, no. A97-98, pp. 548-556, 2002.

3. W. Claes, W. Sansen and R. Puers, "A 40 μA/channel compensated 18-channel strain-gauge measurement system for stress monitoring in dental implants", *IEEE Journal of Solid-State Circuits*, vol. SC-37, no. 3, pp. 293-301, March 2002.

4. R. Puers, W. Claes, W. Sansen, M. De Cooman, J. Duyck and I. Naert, "Towards the limits in detecting low-level strain with multiple piezo-resistive sensors", *Sensors and Actuators*, no. A85, pp. 395-401, 2000.

5. W. Claes, R. Puers and W. Sansen, "Low-power interfacing for piezo-resistive sensors", *Journal A*, vol. 40, no. 1, pp. 26-30, 1999.

6. R. Puers, P. Wouters and W. Claes, "Remote sensing of temperature, pressure or acceleration: the need for adaptable interface circuits", *Journal A*, vol. 40, no. 1, pp. 12-19, 1999.

Conference papers

1. W. Claes, M. De Cooman, W. Sansen and R. Puers, "A 136-μW/channel autonomous strain-gauge datalogger", *IEEE International Solid-State Circuits Conference (ISSCC), Digest of technical papers*, pp. 196-197, San Fransisco, California, February 9-13, 2003.

2. R. Puers, W. Claes, W. Sansen, M. De Cooman, J. Duyck and I. Naert, "A miniaturized, autonomous, programmable stress monitoring device, part of a dental prosthesis", *Proceedings of Tranducers'01/Eurosensors XV*, pp. 52-55, Munich, Germany, June 10-14, 2001.

3. W. Claes, W. Sansen and R. Puers, "A 40 μA/channel compensated 18-channel strain gauge measurement system for stress monitoring in dental implants", *Proceedings of IEEE Custom Integrated Circuits Conference (CICC)*, pp. 101-104, San Diego, California, May 6-9, 2001.

4. W. Claes, R. Puers, W. Sansen, M. De Cooman, J. Duyck and I. Naert, "An implantable, remotely programmable datalogger for stress monitoring in dental implants", *Proceedings of International Symposium on Biotelemetry XVI*, p.33, Vienna, Austria, May 6-11, 2001.

5. W. Claes, W. Sansen and R. Puers, "Evolution in Miniaturized Autonomous Data Loggers for Biomedical Applications", *World Congress on Medical Physics and Biomedical Engineering (WC 2000)*, CDROM, Chicago, Illinois, July 23-28, 2000.

6. W. Claes, R. Puers, W. Sansen, M. De Cooman, R. Geers, H. Van Houten, G. Brunninkhuis, J. Cop and H. Hofman, "Miniaturized telemetric data acquisition devices: the long road from prototype development to market introduction", *Proceedings of International Symposium on Biotelemetry XV*, p. 54, Juneau, Alaska, May 9-14, 1999.

7. R. Puers, W. Claes, W. Sansen, M. De Cooman, J. Duyck and I. Naert, "Towards the limits in detecting low level strain with multiple piezo-resistive sensors", *Proceedings of Eurosensors XIII*, pp. 49-52, The Hague, The Netherlands, September 12-15, 1999.

Papers in workshop proceedings

1. R. Puers, W. Claes, W. Sansen, M. De Cooman, J. Duyck, I. Naert, "A Wireless Programmable Datalogger For Stress Monitoring In Dental prostheses", *Proceedings of 2nd Belgian Day on Biomedical Engineering*, Brussels, Belgium, October 18, 2002.

Articles in book of articles

1. W. Claes, R. Puers, W. Sansen, M. De Cooman, R. Geers, H. Van Houten, G. Brunninkhuis, J. Cop and H. Hofman, "Miniaturized telemetric data acquisition devices: the long road from prototype development to market introduction", article in book entitled *Biotelemetry*, edited by International Society on Biotelemetry, pp. 569-579, Wageningen, The Netherlands, 2000.

Bibliography

[3M] www.3m.com, *Test and burn-in grid ZIP sockets*, Datasheet.

[Abi 83] A. Abidi and R. Meyer, "Noise in Relaxation Oscillators", *IEEE Journal of Solid-State Circuits*, vol. 18, no. 6, pp. 794–802, Dec. 1983.

[Abr 72] M. Abramowitz and I. A. Stegun, Eds., *Handbook of Mathematical Functions with Formulas, Graphs, and Mathematical Tables, 9th printing*, p. 14, New York: Dover, 1972.

[Alc 97] Alcatel Microelectronics, *Model cards BSIM3v3 C07MA and C07MD*, 1997.

[Alc 01] Alcatel Microelectronics, *Electrical parameters, DS 13291*, 2001.

[Alp] http://www.alpha elec.co.jp, *Ultra-Precision Chip Resistor (Wire Bonded)*, Product Data Sheet.

[Alt] www.altera.com, *ByteBlaster parallel port download cable*, Datasheet, Altera, San Jose, California, USA.

[Avo] www.avonwood.co.uk, *Eureka 411 Tag Industrial Decoder*, Datasheet, Avonwood Developments Ltd., Wimborne, UK.

[Bas 98] J. Bastos, *Characterization of MOS Transistor Mismatch for Analog Design*, PhD thesis, ESAT-MICAS, K.U.Leuven, Belgium, Apr. 1998.

[Beg 97] J. Begueret, M. Benbrahim, Z. Li, F. Rodes, and J. Dom, "Converters dedicated to long-term monitoring of strain gauge transducers", *IEEE Journal of Solid-State Circuits*, vol. 32, no. 3, pp. 349–356, Mar. 1997.

[Ber 88] G. Bergmann, F. Graichen, and A. Rohlmann, "Multichannel strain gauge telemetry for orthopaedic implants", *Journal of Biomechanics*, vol. 21, pp. 169–176, 1988.

[Bev 69] P. Bevington, *Data Reduction and Error Analysis for the Physical Sciences*, pp. 58–64, New York: McGraw-Hill, 1969.

[Bol 95] D. Bolliger, Ed., *Integration of an ultraviolet sensitive flame detector*, ETH Zurich, 1995.

[Bow 86] L. Bowman and J. Meindl, "The packaging of implantable integrated sensors", *IEEE Transactions on Biomedical Engineering*, vol. 33, no. 2, pp. 248–255, Feb. 1986.

[Bru] www.bruco.nl.

[BSI] www.brilliancesemi.com, *BS62LV2000*, Datasheet, Brilliance Semiconductor Inc., Irvine, California, USA.

[Cap 96] P. Cappa, Z. D. Prete, and F. Marinozzi, "Experimental analysis of a new strain-gage signal conditioner based on a constant-current method", *Sensors and Actuators*, vol. A55, pp. 173–178, 1996.

[Coz 99] A. Cozma, *Anodic bonding for the fabrication of self-testable pressure sensors*, PhD thesis, ESAT-MICAS, K.U.Leuven, Belgium, 1999.

[Deb 02] D. D. Bruycker, *Intelligent silicon pressure sensors*, PhD thesis, ESAT-MICAS, K.U.Leuven, Belgium, 2002.

[Dup a] www.dupont.com, *5091D/5092D/5093D PTC Thermistor Compositions*, Product Data Sheet.

[Dup b] www.dupont.com, *Birox 2000 series resistors*, Technical information.

[Duy 00] J. Duyck, *Biomechanical characterisation of in vivo load on oral implants*, PhD thesis, Department of Prosthetic Dentistry, K.U.Leuven, Belgium, 2000.

[Ele] www.electroscience.com, *NTC-2100 series*, Datasheet.

[Ene] www.energizer.com, *Energizer no. 392*, Engineering Datasheet.

[Enz 87] C. Enz, E. Vittoz, and F. Krummenacher, "A CMOS Chopper Amplifier", *IEEE Journal of Solid-State Circuits*, vol. 22, pp. 335–342, June 1987.

[Enz 96] C. Enz and G. Temes, "Circuit Techniques for Reducing the Effects of Op-Amp Imperfections: Autozeroing, Correlated Double Sampling, and Chopper Stabilization", *Proceedings of the IEEE*, vol. 84, no. 11, pp. 1584–1614, Nov. 1996.

[Esp 98a] M. Esposito, J.-M. Hirsch, U. Lekholm, and P. Thomsen, "Biological factors contributing to failures of osseointegrated oral implants. (I) Success criteria and epidemiology", *European Journal of Oral Science*, vol. 106, no. 1, pp. 527–551, Feb 1998.

[Esp 98b] M. Esposito, J.-M. Hirsch, U. Lekholm, and P. Thomsen, "Biological factors contributing to failures of osseointegrated oral implants. (II) Etiopathogenesis", *European Journal of Oral Science*, vol. 106, no. 3, pp. 721–764, Jun 1998.

[Fat 90] J. Fattaruso, M. D. Wit, G. Warwar, K. Tan, and R. Hester, "Noise in Relaxation Oscillators", *IEEE Journal of Solid-State Circuits*, vol. 25, no. 6, pp. 1550–1561, Dec. 1990.

[Fol 90] D. Follett, "An externally powered six channel strain gauge transcutaneous telemetry system", in *Implantable telemetry in orthopaedics*, G. Bergmann, F. Graichen, and A. Rohlmann, Eds., F.U. Berlin, 1990, pp. 87–92.

[Gan 95] W. F. Ganong, *Review of medical physiology*, Prentice-Hall International, 1995.

[Gee 01] Y. Geerts, *Design of high-performance CMOS delta-sigma A/D converters*, PhD thesis, ESAT-MICAS, K.U.Leuven, Belgium, Dec. 2001.

[Gob 83] C. Gobet and A. Knob, "Noise analysis of switched capacitor networks", *IEEE Transactions on Circuits and Systems*, vol. 30, no. 1, pp. 37–43, Jan. 1983.

[Gul 01] K. Gulati and H. Lee, "A Low-Power Reconfigurable Analog-to-Digital Converter", *IEEE Journal of Solid-State Circuits*, vol. 36, no. 12, pp. 1900–1911, Dec. 2001.

[Hab 97] A. Häberli, Ed., *Compenstation and calibration of IC mirosensors*, ETH Zurich, 1997.

[Hor 97] G. van der Horn and J. Huijsing, "Integrated smart sensor calibration", *Analog Integrated Circuits and Signal Processing*, vol. 14, pp. 207–222, 1997.

[Hos 94] S. Hoshaw, J. Brunski, and G. Cochran, "Mechanical loading of Bränemark implants affects interfacial bone modelling and remodelling", *International Journal of Oral and Maxillofacial Implants*, vol. 9, pp. 345–360, 1994.

[Hos 97a] B. Hosticka, *Integrated sensor systems in CMOS technology*, pp. 219–241, Analog Circuit Design, W. Sansen, J. H. Huijsing, R. J. van de Plassche (eds.) Kluwer Academic Publishers, 1997.

[Hos 97b] B. Hosticka, W. Brockherde, and D. Hammerschmidt, "Silicon sensor systems", *Analog Integrated Circuits and Signal Processing*, vol. 14, pp. 261–273, 1997.

[Hui 92] J. Huijsing, "Integrated smart sensors", *Sensors and Actuators*, vol. A30, pp. 167–174, 1992.

[Hui 94] J. Huijsing, F. Riedijk, and G. van der Horn, "Developments in integrated smart sensors", *Sensors and Actuators*, vol. A43, pp. 267–288, 1994.

[Ing 97] M. Ingels and M. Steyaert, "Design Strategies and Decoupling Techniques for Reducing the Effects of Electrical Interference in Mixed-Mode IC's", *IEEE Journal of Solid-State Circuits*, vol. 32, no. 7, pp. 1136–1141, July 1997.

[Ing 00] M. Ingels, *CMOS interface circuits for optical communication*, PhD thesis, ESAT-MICAS, K.U.Leuven, Belgium, Mar. 2000.

[Joh 97] D. A. Johns and K. Martin, *Analog Integrated Circuit Design*, John Wiley & Sons Ltd., 1997.

[Lak 88] K. Lakshimikumar, M. Copeland, and R. Hadaway, "Reply to 'A comment on: Characterization and modeling of mismatch in MOS transistors for precision analog design"', *IEEE Journal of Solid-State Circuits*, vol. 23, no. 1, pp. 296, Feb. 1988.

[Lee 85] K.-L. Lee and R. Meyer, "Low-distortion switched-capacitor filter design techniques", *IEEE Journal of Solid-State Circuits*, vol. 20, no. 6, pp. 1103–1112, Dec. 1985.

[Lem 93] C. Leme, Ed., *Oversampled interfaces for IC sensors*, ETH Zurich, 1993.

[Lie 99] S. Lievens, "Meerkanaals rek-meetsysteem voor prothetische tandheelkunde", Thesis Biomedische en Klinische Ingenieurstechnieken, K.U.Leuven, 1999.

[Mal 96] P. Malcovati, Ed., *CMOS thermoelectric sensor interfaces*, ETH Zurich, 1996.

[Mar 82] K. Martin, "New clock feedthrough cancellation technique for analogue MOS switched-capacitor circuits", *Electronics Letters*, vol. 18, no. 1, pp. 39–40, Jan. 1982.

[Mar 99] A. Marques, *High Speed CMOS Data Converters*, PhD thesis, ESAT-MICAS, K.U.Leuven, Belgium, Jan. 1999.

[Mat 87] H. Matsumoto and K. Watanabe, "Spike-Free Switched-Capacitor Circuits", *Electronics Letters*, vol. 23, no. 8, pp. 428–429, Apr. 1987.

[McC 75] J. McCreary and P. Gray, "All-MOS Charge Redistribution A/D Conversion Technique-Part I", *IEEE Journal of Solid-State Circuits*, vol. 10, no. 6, pp. 371–379, Dec. 1975.

[Mei 94] G. Meijer, "Concepts and focus point for intelligent sensor systems", *Sensors and Actuators*, vol. A41-42, pp. 183–191, 1994.

[Men 97] C. Menolfi and Q. Huang, "A Low-Noise CMOS Instrumentation Amplifier for Thermoelectric Infrared Detectors", *IEEE Journal of Solid-State Circuits*, vol. 32, pp. 968–976, July 1997.

[Men 98] C. Menolfi and Q. Huang, "A CMOS instrumentation amplifier with 600nV offset, 8.5 nV/$\sqrt{\text{Hz}}$ noise and 150dB CMRR", in *Proceedings Custom Integrated Circuits Conference*, 1998, pp. 369–372.

[Mid 89] S. Middelhoek and S. Steigerwald-Audet, *Silicon sensors*, Academic Press, London, 1989.

[Mok 99] W. Mokwa and U. Schnakenberg, "Implantable microdevices", in *Proceedings Eurosensors XIII*, Sept. 1999, pp. 741–746.

[Mom 94] A. Mombelli, "Criteria for success. Monitoring", in *Proceeedings of the first European Workshop on Periodontology*, N. P. Lang and T. Karring, Eds., Quintessence, Berlin, 1994, pp. 317–325.

[Mur 00] T. Murphy, *Autozero amps give extra boost*, www.eetimes.com, 2000.

[Nag 01] M. Nagy, C. Apanius, J. Siekkinen, and H. Estrada, "A user-friendly, high-sensitivity strain gauge", *Sensors*, vol. 18, no. 6, Jun 2001.

[New 01] B. H. Newman and C. A. Martin, "The effect of hot beverages, cold beverages, and chewing gum on oral temperature", *Transfusion*, vol. 41, no. 10, pp. 1241–1243, Oct 2001.

[Nov] www.nova eng.com, *Constellation E Development System*, Datasheet, Nova Engineering Inc., Cincinnati, Ohio, USA.

[Oos 00] H. Van Oosterwyck, *Study of biomechanical determinants of bone adaptation around functionally loaded oral implants*, PhD thesis, Department of Mechanics, Division of Biomechanics and Engineering Design, K.U.Leuven, Belgium, May 2000.

[Opt 86] F. Op't Eynde, "Stroomreferenties en Spanningsreferenties", Thesis Burgerlijk Elektrotechnisch Ingenieur, K.U.Leuven, 1986.

[Pel 89] M. Pelgrom, A. Duinmaijer, and A. Welbers, "Matching properties of MOS transistors", *IEEE Journal of Solid-State Circuits*, vol. 24, no. 3, pp. 1433–1439, Oct. 1989.

[Pet 86] P. V. Peteghem, *Accuracy and resolution of switched capacitor circuits in MOS technology*, PhD thesis, ESAT-MICAS, K.U.Leuven, Belgium, 1986.

[Phy a] www.physikinstrumente.com, *E-662.SR*, Datasheet.

[Phy b] www.physikinstrumente.com, *P-841.40*, Datasheet.

[Phy c] www.physikinstrumente.com, *P-841.60*, Datasheet.

[Pie 98] T. Piessens and J. V. den Keybus, "Ontwerp van een sensor interface chip voor animal monitoring", Thesis Burgerlijk Elektrotechnisch Ingenieur, K.U.Leuven, 1998.

[Pol 94] S. Tsyma, F. van Herwijnen, and P. Leijendeckers, Eds., *Polytechnisch zakboekje*, Koninklijke PBNA Den Haag, 1994.

[Pue 93] R. Puers, "Converting stress into strain: basic techniques", in *Monitoring of orthopedic implants*, F. Burny and R. Puers, Eds., Amsterdam, The Netherlands, 1993, pp. 33–53.

[Pue 96] R. Puers, "Linking sensors with telemetry: impact on the system design", *Sensors and Actuators*, vol. A52, pp. 169–174, 1996.

[Pue 99] R. Puers, P. Wouters, and W. Claes, "Remote sensing of temperature, pressure or acceleration: the need for adaptable interface circuits", *Journal A*, vol. 40, no. 1, pp. 12–19, 1999.

[Rab 96] J. Rabaey, *Digital Integrated Circuits — A Design Perspective*, Prentice-Hall, Inc., Englewood Cliffs, NJ, 1996.

[Raz 95] B. Razavi, *Principles of Data Conversion System Design*, IEEE Press, 1995.

[Raz 01] B. Razavi, *Design of Analog CMOS Integrated Circuits*, McGraw-Hill, New York, 2001.

[Red] www.redlion controls.com, *Model PAX - 1/8 Din Analog Input Panel Meters*, Datasheet.

[Rey] S. Reyntjens and R. Puers, *Safeconequip Project*, Internal document.

[Rey 02] S. Reyntjens, *Three-dimensional micromachining for microelectromechanical systems (MEMS)*, PhD thesis, ESAT-MICAS, K.U.Leuven, Belgium, Oct. 2002.

[Rie 94] F. Riedijk and J. Huijsing, *Indirect converters and oversampling for application in monolithic smart sensors*, pp. 205–222, Analog Circuit Design, W. Sansen, J. H. Huijsing, R. J. van de Plassche (eds.) Kluwer Academic Publishers, 1994.

[Roa 75] R. Roark and W. Young, *Formulas for stress and strain*, Tokio: McGraw-Hill, 1975.

[Sak 83] T. Sakurai and K. Tamaru, "Simple Formulas for Two-and Three-Dimensional Capacitances", *IEEE transactions on electron devices*, vol. 30, no. 2, pp. 183–185, Feb. 1983.

[San 82] W. Sansen, "On the integration of an internal human conditioning system (Invited Paper)", *IEEE Journal of Solid-State Circuits*, vol. 17, pp. 513–521, June 1982.

[San 94] W. Sansen and K. Laker, *Design of Analog Integrated Circuits and Systems*, McGraw-Hill, New York, 1994.

[Sua 75] R. Suarez, P. Gray, and D. Hodges, "All-MOS Charge Redistribution A/D Conversion Technique-Part II", *IEEE Journal of Solid-State Circuits*, vol. 10, no. 6, pp. 379–385, Dec. 1975.

[Sun 02] M. Sund-Levander, C. Forsberg, and L. K. Wahren, "Normal oral, rectal, tympanic and axillary body temperature in adult men and women: a systematic literature review", *Scandinavian Journal of Caring Science*, vol. 16, no. 2, pp. 123–128, Jun 2002.

[Van 00] A. Van den Bosch, M. Steyaert, and W. Sansen, "An accurate statistical yield model for CMOS current-steering D/A converters", in *Proceedings IEEE International Symposium on Circuits and Systems*, Geneva, May 2000, pp. IV105–108.

[Vdb 02] J. Vandenbussche, *Systematic design of analog IP blocks*, PhD thesis, ESAT-MICAS, K.U.Leuven, Belgium, Oct. 2002.

[VdP 01] G. V. der Plas, *A computer-aided design and synthesis environment for analog integrated circuits*, PhD thesis, ESAT-MICAS, K.U.Leuven, Belgium, Dec. 2001.

[Wec 02] R.Wechsung, "Market analysis for microsystems 2000-2005 - A report from the NEXUS task force", *MST News*, , no. 2/02, pp. 43–44, Apr. 2002.

[Weg 87] G. Wegmann, E. Vittoz, and F. Rahali, "Charge injection in analog MOS switches", *IEEE Journal of Solid-State Circuits*, vol. 22, no. 6, pp. 1091–1097, Dec. 1987.

[Wer 00] M. Werner, S. Heyn, and T. Kölner, "Microsystems in health care markets: recent achievements and future trends", *MST News*, , no. 4/00, pp. 4–7, Sept. 2000.

[Wis 94] K. Wise, *Analog data acquisition circuits in integrated sensing systems*, pp. 117–140, Analog Circuit Design, W. Sansen, J. H. Huijsing, R. J. van de Plassche (eds.) Kluwer Academic Publishers, 1994.

[Wou 95] P. Wouters, *Intelligent and low power sensor interface systems for biotelemetry applications*, PhD thesis, ESAT-MICAS, K.U.Leuven, Belgium, Apr. 1995.

[Yin 92] G. Yin, F. Op't Eynde, and W. Sansen, "A High-Speed CMOS Comparator with 8-b Resolution", *IEEE Journal of Solid-State Circuits*, vol. 27, no. 2, pp. 208–211, Feb. 1992.

[Zia 97] B. Ziaie, M. Nardin, and A. Coghlan, "A single-channel implantable microstimulator for functional neuromuscular stimulation", *IEEE Transactions on Biomedical Engineering*, vol. 44, no. 10, pp. 909–920, Oct. 1997.

Index